ECOLOGICAL ENTANGLEMENTS

This book shows us some ways by which we can recover the human subject even in the pursuit of evidentiary, justified systems of knowing, being and living. It offers insights into possible paradigms through which we can eventually produce not just a 'Green Ecology' but also a 'Green Biology'.

—**Sundar Sarukkai** (from 'Foreword'), *Founder*, *Barefoot Philosophers*

ECOLOGICAL ENTANGLEMENTS

AFFECT, EMBODIMENT AND ETHICS OF CARE

Edited by

Ambika Aiyadurai
Arka Chattopadhyay
Nishaant Choksi

Orient BlackSwan

ECOLOGICAL ENTANGLEMENTS:
AFFECT, EMBODIMENT AND ETHICS OF CARE

ORIENT BLACKSWAN PRIVATE LIMITED

Registered Office
3-6-752 Himayatnagar, Hyderabad 500 029, Telangana, India
e-mail: centraloffice@orientblackswan.com

Other Offices
Bengaluru, Chennai, Guwahati, Hyderabad, Kolkata,
Mumbai, New Delhi, Noida, Patna

First published 2023

041025

ISBN 978-93-5442-259-1

Typeset in Calisto MT 10/12 *by*
Shine Graphics, Delhi 110 094

Printed at
Avantika Printers Private Limited, New Delhi 110 020

Published by
Orient Blackswan Private Limited
3-6-752, Himayatnagar, Hyderabad 500 029, Telangana, India
e-mail: info@orientblackswan.com

CONTENTS

FOREWORD
GREENING THE FRONTIERS OF ECOLOGY

Sundar Sarukkai

During the period in which I was writing this Foreword, I was also seeing a remarkable Netflix series called *Green Frontier*. It is a moody, experiential look not just at the mysteries of the Amazon forest, but also at the consequence of forgetting the mysteries of the human–nature relation. While there may be questions on the exact representation of Indigenous tribes in this series, it is nevertheless a show that articulates very important questions on alternate knowledge systems of nature and health (in contrast to 'Western' modernist paradigms) as well as different conceptualisations of both nature and human. The world is populated with such diverse understandings of nature and human, and yet ecology as a discipline is still built on the foundations of some dominant metaphors related to a limited understanding of what terms like 'nature', 'human' and 'science' could be. In this spirit, I find that this volume can be understood as trying to build a green frontier to the discipline by showing how the boundaries and borders of disciplines that deal with nature in all its multiplicity cannot be encroached upon or taken over by hegemonic and dominant cultural narratives, especially since these approaches are primarily utilitarian in nature.

Although this is a book about ecology, it is raising—through its varied content—questions about the foundations of scientific analysis and what it means to be a science. It focuses on one of the most important blind spots in scientific methodology, namely, its silence about the role of affect, morality, memory and beliefs in scientific practice as well as scientific knowledge. This silence can be read in different ways: one, it could only be an (epistemological?) inability to factor essential elements of the human scientist–subject; or two, it could be a recognition that such a move goes against certain ideological claims about the nature of science and scientific methods. However, the presence of the human subject along with its desirable and undesirable qualities cannot be erased from science, both in practice and in its discourse. Attempts to ignore the elephant in the room is truly another 'ecological' problem!

This book brings together reflections from different disciplines and perspectives, focusing on questions of affect, body, embodiment, care, language and the human–nonhuman interaction. In so doing, it attempts to present the discourse of ecology in its varied forms. An integral part of the scientification of any discipline is to compartmentalise and reduce complexity to forms that are amenable to analysis. Ecology as a discipline has also undergone these discursive surgeries. But in so doing, many aspects of the conceptual world of ecology are either left out or become marginalised. Typically in any science, what gets removed from the scientification of a discipline are the human, social and cultural aspects of the practice and knowledge produced in that discipline. In the case of ecology, these aspects are not marginal in any sense and this book attempts to integrate these dimensions into the mainstream practice of this discipline.

There are deep implications of such integration. The point that interests the editors of this book is that such integrative practices lead to an epistemology which is not only about knowing the other but also about caring for the other. But in stating it in this manner, they immediately enter a contentious domain, that of challenging the aims and goals of any science. Can caring be a legitimate scientific goal? Whether this book can answer this to the readers' satisfaction is not as important as recognising that in taking this step, this book is already embarking on an important journey.

It is not enough to say that the effort of a discipline is not just in producing knowledge but also in talking about care and caring for the other. For this goal to be accomplished, like for any goal, certain fundamental conceptual changes have to be made. This book attempts to make some such conceptual reordering such as questioning the binary of the human and the nonhuman. Rather than locate these two categories in opposition, or as one specified through differences, they begin with understanding the interaction between them. This is an interesting philosophical approach of focusing not on entities but on the relations between entities, or equivalently focusing on a process philosophy. This is a difficult task since it has been a common project of both philosophy and science to distinguish humans from nonhumans. This attitude gets so ingrained in intellectual practices that cultures begin to be distinguished—as in colonial discourse—in ways that are parallel to the human–animal distinction. For example, rationality becomes a marker to distinguish humans and animals, and the same distinction is made between colonising European cultures and the colonised cultures of Asia and Africa. So, the task of dismantling the hierarchy between human and nonhuman is not easy for it challenges the very foundations of science. In raising this challenge, the book has the potential to question the scientificity of ecology, or in other words, to ask the question 'In what sense is ecology truly a science?' And when this question is raised, what is being asked is not whether ecology matches the expectation of what a science should be (as defined by sciences like physics), but more importantly, what should be added to the elements of science in order to fulfil the potential of ecology as a discipline.

The second major conceptual change which is required is more a discursive step—one that challenges the history of disciplines. When the editors of this volume point out that Asian and African cultures have had a long history which did not accept a facile distinction between humans and nonhumans and that any rethinking of this binary might find useful resources in these earlier traditions, they are challenging the received notion that 'modern' science and scientific ideas must be derived only from certain cultural contexts, such as Europe from the sixteenth or seventeenth century onwards. The refusal of modern science to acknowledge potentially important conceptual ideas from earlier cultural practices in Asia and Africa (with the exclusion of the Greeks who are artificially constructed as the progenitors of science in Europe) is an obstacle to this vision of drawing upon conceptualisations of human and nonhuman within a global science. Whether or not the physical sciences can find resources in non-Western cultures is not the relevant question here since in the rich conceptualisations of nature, in definitions of human and in the complex narratives on the relationship between humans and nature/nonhuman, there is a huge surplus of ideas and approaches present in non-Western traditions. Can ecology as a discipline continue to mimic the exclusionary practice of the other sciences in keeping the knowledges of the non-West out of the purview of the everyday practices of this discipline?

The third major conceptual challenge raised by this book is to relook at the question of ethics as an essential component of scientific knowledge. The editors note that to 'understand ecology is to understand the ethical relations among different forms of life' (p. 2, in this volume). What are the implications of this statement? Ethics is another domain which captures the presence of the human subject. There is a long history of how modern science finds ways to remove questions of ethics out of the purview of its practice and epistemology. An example of this is the way curiosity comes to occupy such a central position in scientific practice.[1] Historically, the rehabilitation of curiosity made possible the explicit divorce of ethical considerations in science. The fundamental reason why ethics is a problem for science is again related to the presence of the human subject in scientific articulations. If scientific knowledge is seen as knowledge of the world independent of human influence and human interests, then morality—seen as an integral expression of what it is to be human—must also be removed from its discourse. So, if ecology is going to essentially incorporate ethical relations as part of its disciplinary practice and discourse, then it has to challenge certain fundamental assumptions about science itself.

Another conceptual challenge such an approach forces on the idea of science is the focus on affect and expression. Affect plays no role in scientific knowledge. In fact, the lengths to which science goes to remove affect is illustrated by the stringent rules governing scientific writing. The absence of any marker of emotion even when writing about a great and surprising discovery is a form of institutionally removing any presence of affect, because affect too is seen as a defining marker of being human. So what does it mean for ecology as a science to include affect as part of its disciplinary practice? How will this be accomplished? What happens to the idea of 'objective' knowledge which historically has come to mean something in opposition to products of emotion? The strongly entrenched binary of reason and emotion captures this fundamental tension operating within science. So how does ecology retain its role as a science when it wants to consider affect as a part of its concerns?

The emphasis on embodiment extends this question since the body is the first absentee in science. The best way to remove the human out of the equations of science is to remove any meaningful function of the body. Philosophically, the attempts to remove secondary qualities (those that arise in interactions with the body, such as taste) is already a good indication of how the problem of the body has to be dealt with in any discipline that aspires to be called a science. The removal of the body is also the removal of gender as well as cultural and social characteristics. The universality of science is very dependent on these moves related to the erasure of the human subject. So, when this new ecology desires to 'embody itself', what happens to its status as a science and what happens to the justification of the knowledge it produces? We should note that even social science—in imitating the natural sciences—struggles with including elements such as affect, embodiment, care and so on, as part of its legitimate discursive structure.

Why is something so obvious and at the same time so difficult to understand or describe? Disciplines open up spaces to understand their themes/concepts, but they also foreclose them in many ways. Putting a concept within a discipline is to put a boundary around that concept. One of the most entrenched fences around concepts in the sciences is that of talking about affect or anything that is remotely related to the presence of the human subject. The trouble for science is the human subject and scientific methodology can be seen historically as an attempt to find ways to erase the hovering presence of the subject. After all, it is the human scientist filled with cultural and ideological, subjective beliefs that produces truths about nature. The human individual—in their cognitive splendour as well as in their affectual behaviour—is the discoverer of truths of the world. The inherent tension between discovery and invention is a testament to the tension in acknowledging the human in the room.

While disciplines like physics and mathematics can find ways to proceed while leaving the human behind, it is not so easy with biology. The human is after all the fulcrum, the tensile point, around which biology has to be constructed. Biology can remove the presence of the subject by removing the most difficult characteristics of the subject but in so doing, does it lose certain capacities for a more encompassing understanding of its own subject matter? When we ask this question to ecology, equally difficult problems arise. We should recognise that even an everyday concept of 'nature' is always incompletely described in different subjects. The question we have to ask is, which nature? Nature as it is defined for physics? Or for biology? Or for psychology? Or for mathematics? How does ecology construct its concept of nature that allows it to engage with the concerns of the field without being completely bullied by some naive idea of scientific method?

This book shows us some ways by which we can recover the human subject even in the pursuit of evidentiary, justified systems of knowing, being and living. It offers insights into possible paradigms through which we can eventually produce not just a 'Green Ecology' but also a 'Green Biology'.

Note

1. See Sundar Sarukkai, "Science and the Ethics of Curiosity", *Current Science*, vol. 97, no. 6, 2009.

ACKNOWLEDGEMENTS

We would like to thank the Indian Council of Social Science Research (ICSSR) and the Department of Humanities and Social Sciences, IIT Gandhinagar for helping us organise the 2020 (March 2–4) international conference, 'Affect, Embodiment and Ecology: Multidisciplinary Perspectives' that germinated this volume of essays.

PUBLISHER'S ACKNOWLEDGEMENTS

Fig 15.1 from *Boi Sangraha 2* by Subimal Mishra published by Gangchil. Reproduced with permission of the publishers.

INTRODUCTION

Ambika Aiyadurai, Arka Chattopadhyay & Nishaant Choksi

The condition of human and nonhuman species is of grave concern as the planet continues to face unprecedented environmental damage. Calling for new ways to apprehend the ecological crisis, this book seeks to formulate a framework that integrates the social, material and cultural dimensions of ecology. This conception comprises not only relations among different forms of life but also an examination of affective connections and shared spaces that co-construct a sense of 'being–together'. Therefore, the volume argues that ecology, both as a discursive and material form, is thoroughly entangled with affective, communicative and embodied practices that provide a sense for not only understanding, but also caring for the other. This theoretical and methodological move is even more necessary both in times of anthropogenic climate change and virogenic pandemics.

These issues have previously been addressed from particular disciplinary vantage points and from within specific disciplinary enclosures. This volume generates a set of cross-disciplinary conversations from fields as diverse as history, anthropology, ethology, linguistics and literary studies through an overarching conception of ecology. The aim of the volume is to create multiple lines of inquiry that bring together discussions of affect, embodiment and ethics. In doing so, the book makes theoretical and methodological interventions within the respective disciplines, while also laying the groundwork for a better addressal of contemporary ecological crises.

Our notion of ecology rests on the fact that the interactional encounters between humans and nonhumans demonstrate a complex multispecies intersubjectivity in which all lives (*zoe* rather than *bios*) and cultures are enmeshed. The simplistic dichotomy of human and nonhuman spheres is increasingly becoming an obsolete framework to understand our ecological crisis. Scholars, therefore, are calling for a 'post-anthropocentric' or 'anthro-de-centered orientation' to re-situate the 'anthropos' (Küpers 3). Increasingly, human and nonhuman lives are becoming biologically, culturally and politically intertwined. The book thus intends to study these connections between human and nonhuman that are always already co-constituted and mediated through bodies. As Aisher and Damodaran argue, 'human ecological relations unfold historically through "meaningful exchange among human and nonhuman entities"' (298). Their notion, however, of 'exchange' misses the extending practices of care to nonhuman Others (Kohn; Tsing). This book asks what it means to care for others, and how an ethic of care can emerge from interactional and affective encounters within embodied spaces. To understand ecology is to understand the ethical relations among different forms of life. This ecological being–together requires an analysis of how multispecies care can be positioned in relation to ethical encounters that occur within ecological spaces. It opens us to the affective possibilities of nurture and care for nonhuman species.

In a well-known formulation, Brian Massumi discusses affect as situated bodily feelings that are 'autonomous to the degree to which it escapes confinement in the particular body whose vitality, or potential for interaction, it is' (96). For Massumi, affects are thus 'autonomous', 'open' and 'virtual'. Affects, however, are also intensely ecological in their outwardness, invoking questions that move us beyond Massumi's conception. For instance, the volume poses questions such as: how does the human and animal body express and situate itself? If affect gets inscribed on the body and the body acts and reacts in relation to it in particular ways, does this affective embodiment have a link with larger ecological spaces?

Moreover, to view affect as not autonomous but ecological, raises the question of how language is entangled with the human and nonhuman worlds. Linguistic and cognitive theories have long placed the knowing subject ('competence') within the mind of individuals rather than in their interaction with bodies, affects and environments ('performance') (Chomsky). The book rethinks the meaning of competence as an affectively charged and embodied mode of cognition that is intimately entangled with communicative ecology. New models of 'communicative competence' 'incorporate context into our bodies and minds and mould both bodies and contexts together' through our multi-modal, intersubjective communicative practices (Kataoka et al.). Communication is then less a 'meeting of the minds' and more of an experience of 'world-making' (Streeck) through the coordination of socially indexical gestures and speech, much like a dance or a sport.

Finally, along with affect, care and language, we also open up the question of embodiment as a critical component for the conception of ecology. In a human–animal ecology we could consider the animal body as a corporeal surface on which human beings often project anthropomorphic feelings, humanising the animal body in the process. Whether they project these feelings or not, nonhuman animal bodies interact with human bodies through signals. An iconic example of the animal's reaction to the human is the monkey's near-perfect mimicking of a human gesture seen moments ago. This projection, interaction and imitation produce a dialogic space of entangled embodiments. The neurobiological signalling that happens through mirror neurons has implications for an ecological understanding of embodied space that goes beyond the human. As Louise Barrett observes, 'actions produced by mechanical objects but otherwise identical to those performed by a human or a monkey do not trigger a mirror response, only those from another animal' (35). This means there is something particular about this animal mirroring. It forms not only a neurocognitive but a social ecology as well with larger underpinnings.

Setha Low defines 'embodied space' as a multidisciplinary construction in the anthropological process of 'place-making' by way of 'a focus on bodies as they create space through mobility and movement' (20). We could consider the ecological concentration of body and space in Australian aboriginal communities and their indigenous map of the land in which each body of elders (past and present), animals and insects—not to mention inanimate objects—co-constitute a collective embodied space. Low's study (conducted in the 1980s and 90s), on walking pace and patterns across different gender and age groups in two plazas of Costa Rica, draws on behavioural maps to establish the human body as a crucial agent in the social construction of space.

Embodied space is not without its political implications. For instance, we could mention Swati Chattopadhyay's reading of political wall writing in the Indian city of Kolkata in colonial and postcolonial timelines. She foregrounds the political nature of this embodied act of place-making and draws our attention to how political posters on the walls led to corporal punishment from the repressive state apparatus (57). In a different politicisation of embodied space, Aniket Jaaware in *Practicing Caste* closely attends to the spatial register of touchability and untouchability as a social construction that underlines how caste is performed in India. This once again requires a notion of embodied space and Jaaware goes deep into social prohibitions such as touching the sexual organ in public space. He extracts a phenomenology of caste practice from the embodied space of touching and not touching. It goes back to the origins of these corporeal practices in a child's development and space is key to this genealogy. Be it caste or political action, spatial embodiment is an important notion. When we analytically construct an ecology in which affective communication, embodied space and ethics of caring are crucial, it is important to be true to the affordances and entanglements that enable this non-hierarchised, flat and relational ontology of ethos.

In this attempt to think through such an eco-ethology, the book divides the contributions into three major sections: 'Decentering Humans and Problematising Care'; 'Ecological Affordances and Affective Expressions' and 'Embodied Spaces and Ecological Entanglements'.

DECENTERING HUMANS AND PROBLEMATISING CARE

In Indigenous philosophies, the lives of humans and nonhumans overlap, preventing the dominance of a human-centric society. These non-Western societies consider themselves part of an extended ecological family, fostering kin relations and assigning personhood status to nonhuman beings (Morris). Therefore, this is a process of 'becoming with' others who share the planet (Harraway). Such an integrated worldview is largely absent in the Western industrialised world, where nature is just a 'resource' to be exploited for human utility. The severe environmental crisis and current pandemic has been calling for a paradigm shift in the way the modern world perceives nature, including caring for nature. Usually conceived of as benevolent, care may also become coercive (van Dooren, "Care"). Certain forms of care could produce undesired and uncomfortable results leading to the 'violent-care of conservation' (van Dooren, "A Day with Crows").

This section of the volume rethinks the human and nonhuman through an embodied nexus of society, ecology, and culture. Exploring the 'shared worlds' inhabited by both animals and humans, "Ethos, Pathos, Logos: Affective and Emotive Ethnographies of Human–Macaque Lifeworlds" by Ishita Ramakrishna and her colleagues examines the relations of humans with Nicobar long-tailed macaques.

Their rich field work in the Andamans challenges our understanding of the binary terms associated with human–primate relations: conflict and coexistence. This multi-author study traces emotion and affectual geographies of how humans perceive and behave towards primates. Using the three Aristotelian methods of persuasion—ethos, pathos and logos—the authors problematise these relations as strange and complex, having the power to generate a unique individuality in every human–nonhuman relationship. Ecological entanglements are further examined in "How to Become–with a Bird: Lessons from Salim Ali's *Common Bird*" by Krishanunni Hari in his analysis of texts written by the famous ornithologist Salim Ali. The chapter maps how humans and birds appear as emergent subjects of co-constitutive becoming.

In "Feeling the Moss: Exploring Ecological Sensitivity Through 'Enchanted' Encounters", Deborah Datta observes 'enchanted' encounters through nurturing the ecological sensibilities of school students in the dense urban space of Mumbai. Datta argues that such encounters generate a reciprocal sense of mutual care through 'affective and playful labour'. In "Reading the Elephant: Towards Affectual Conceptualisations of the Wild Asian Elephants", Banerjee and Sinha offer an alternative reading of shared elephant–human habitats using Indigenous ethologies, including oral history and the experiential knowledge of local communities. The framework of interdependence between human and nonhuman is also explored by Krishna Kumar in "Blindness and Canine Heroicisation: Interdependence in Kuusisto's *Have Dog, Will Travel*". He reads Stephen Kuusisto's text to delve into the lives of canines who guide visually disabled persons, suggesting a heroicisation of the dog to the estrangement of its human companion. In "Human and Animal: 'Destitution' and 'Divination' in the Discourse of Caste", Ankit Kawade shows how the animal comes to prefigure representations of caste bodies, for instance, in the cow (considered as a mother) with the brahmin and the dog with the untouchables. His paper also discusses a wide range of issues addressing the question of human–animal relations in the discourse of caste in India.

ECOLOGICAL AFFORDANCES AND AFFECTIVE EXPRESSION

A crucial way in which we understand our environment and bring both human and nonhuman others into our life is through language.

While some understand language as a primarily grammatical structure (phonemes, morphemes and so on), there has been a long tradition in linguistics that sees language as 'expressive', comprising much more than referential meaning (Diffloth; Jakobson). The understanding of language as 'expression' rather than 'structure' compels us to rethink the relation between language, affect, and the ecological emplacement of language. Contrary to theorists like Frederick Jameson who argue that affect 'eludes language', the expressive view of language suggests that affect is a multichannel phenomenon' that 'floods linguistic forms on many levels of structure' (Besnier 421). This section will show how the 'affective' element of language, or expressiveness in general is intimately tied to ecology. It suggests that the ecological surround provides a semiotic 'affordance' (Sanders; Windsor) through which language is suffused with affect, and how it then may mediate practices of care and the material conditions of living together.

The section consists of four chapters. Badenoch and Osada's "Expressiveness and Affect: Everyday Poetics in Natural Landscapes" looks at the 'expressive' word class (also called 'ideophones' cf. Dingemanse) among speakers of Indigenous languages in India and Laos to show how affect infuses both the grammatical understanding of language among these speakers as well as natural phenomenon like rain, experiences of moving through forest environments and the relation between human and nonhuman animals. Badenoch and Osada argue that linguistic structure in a language posits an affective intimacy between humans and their ecologies that can be culturally elaborated upon. Nishaant Choksi's "Songs of Script: Embodied Affect and Indigenous Language Literacy in Eastern India" also considers linguistic form, in this case a discussion of script practices among Santali speakers in eastern India. Using a linguistic anthropological approach, Choksi shows how the creation and circulation of the Indigenous Ol-Chiki script in a primarily non-literate community took place by the juxtaposition of script with affective and embodied practices of music and dance. The study demonstrates how the written word emerges as a form of affective expression that transcends its referential relation to spoken language.

The next two chapters highlight the relation between expressiveness, memory, and ecology. Asijit Datta in his "Winding Spools and Speaking Stones: How to Play Beckett's Tape Recorder and Listen to Handke's Stone Woman in *Krapp's Last Tape* and *Till Day You Do Part, or A Question of Light?*" focuses on two plays, Michael Handke's *Till Day*

You Do Part and Samuel Beckett's *Krapp's Last Tape*. He relates to the modernist scepticism of technology as a replacement for the ecological entanglements of affective expression, arguing that the temporal structure of Krapp's recording device actually erases memory and affective engagement, while Handke's petrified 'stone woman' creates a relational subjectivity, becoming an ecologically emplaced speaking subject. In a similar vein, Antara Ghatak's "Embodying the *Birangona*: Negotiating History and Memory of the Bangladesh Liberation War, 1971" explores the testimony of the *birangonas*, or women who were raped and killed during the Bangladesh Liberation War. Discourse, whether that of the Pakistani soldiers or later of the Bangladeshi state, silenced these women creating a lack of any kind of expression. However, drawing on the poetry, narratives and artwork of a famous *birangona* Ferdousi Priyabhashini, Ghatak suggests that it is the engagement with the nonhuman natural world, such as the dead bark of trees—which the *birangona* infuses with new expressivity—aligning her ecological surround with the affective memory of war that previously was only embodied in herself.

EMBODIED SPACES IN ECOLOGICAL ENTANGLEMENTS

From care and affect *vis-á-vis* the environment, we come to an ecological understanding of space in its relationality with human and nonhuman bodies that co-constitutes ecological space. Recent cognitivist understandings of embodiment as a corporeal process have variously highlighted the idea of extension. The embodied mind is extended into the environment as it makes a cognitive coupling with an element from our ecology. It is in this sense that the embodied and extended mind is considered an 'environmental vehicle' (Menary 21). We will follow the footsteps of this embodied environmental vehicle in ecological space. Language, affect and expressions are environmental 'surrounds' for the embodied human subject who in turn interacts with diverse nonhuman organisms. As Naoya Hirose notes, '[t]he ecological approach to embodiment maintains that tool use extends the user's body beyond the surface of the skin' and what it produces is an 'extended body', that is, the body and the tool as a unified system (296). If we map this operation on an environmental notion of space, what we get is an ecological body in space as well as a body that creates space when it interacts with some other human or nonhuman body.

This section of the book will focus on embodied spaces in fiction, social practices, the circus and the historical site of the city in the context of environment and affectivity. Maintaining our transcontinental scope, the section opens with Mark Byron's "Becoming Thylacine: Embodiment, Ecology, and Materialism in Julia Leigh's *The Hunter*" that explores the human relation with ecology in a biotech employee's search for the Tasmanian tiger. At stake is the human body's search for the obsolescent animal body and Byron approaches embodiment through phenomenology as he explores an ethic of 'becoming–animal' in the novel. The chapter addresses the environmental issues of climate change and mass extinction through an embodied quest that is intensely spatial in its concern with the species. Migrating in both space and time from Tasmania to colonial Bengal in India, Prodosh Bhattacharya's "Walking on a Tightrope: Circus and its Double in Colonial Bengal" studies the Bengali circus with its particular affective dynamic of human–animal relations of embodiment in the performative space. The colonial relation of subject–bodies unravels a complex politics in this historical site of performance. The embodied space of the circus here becomes a site of affective events that underpin bodies through and beyond textuality. We stay with colonial Bengal in Anuparna Mukherjee's "Imperial Malady: Empire and Affect in Chronicles of the Raj" that examines the imperial urban space of Calcutta in literary representations through the historical and medicalised affect of nostalgia that comes from comparisons with London, the desire for a portable home and the lived experiences of the coloniser's and colonised bodies. It reads literary and cultural texts on the colony as a spatial double of the colonialist subject's paradise—London.

Antaripa Bharali's "Between Space and the Body: The Affective Economy of Caste" approaches the affective economy of embodied space through the sociological lens of caste practices in India, especially in its corporeal rituals of touching and 'untouching'. It underlines the human body as space by going into the assertion of the dalit body as *dalitvada*. Reading through Ambedkar's reflections on space in the phenomenological operation of the caste system, the chapter argues for space as an important conceptual category with which to consider the practice of caste discrimination. Samrat Sengupta considers the relation between storytelling and narrative world formations in Bengali–Indian literature from the ecological vantage of embodied spaces in "A Sense of Waste: Environment as Alienated 'Setting' and Bengali Experimental Fiction". It engages with themes of urban pollution and climate change in the works of Subimal Mishra by way of a posthumanist approach. The contribution of this chapter lies in building a bridge between the experimental montage-like narrative form of Mishra and the posthumanist ramifications of an anthropogenic climate crisis.

In an attempt to devise an ethic against the backdrop of modernity and climate crisis, Jonathan Lear discusses the visionary native American leader Plenty Coups' dreams of imagining a difficult future of adaptation for the Mountain Crows of Montana. In Coups' dreams, there is an anxiety about their traditional ways of living as the march of urban modernity with climatic devastation risks the very survival of the Crow Nation. But there exists a hope in these dreams that towers over anxiety to generate courage. It is from this point that Lear develops the idea of 'radical hope' that differs from 'mere optimism'. Hoping is a radical action in this formulation and this affect is socially and politically radical. It is charged by the courage to construct a future Crow subjectivity that does not exist in the present. As we have argued throughout this proposal, our book is an affective exercise in entanglements of embodiment, spatiality and ecology in an effort to construct an ethic of caring in an interactive ethos of the human and the nonhuman. Lear reflects thus on the radical nature of hope:

> What makes this hope radical is that it is directed toward a future goodness that transcends the current ability to understand what it is. Radical hope anticipates a good for which those who have the hope as yet lack the appropriate concepts with which to understand it. (103)

Through this book, we hope to arrive at the futurity of developing a new and composite concept of ecology. This new ecological conceptualisation would take into account the inextricable knotting of the human and the nonhuman in embodied spaces and languages, in order to affectively engage in a deep mutual care for the species–being of one another.

Works Cited

Aisher, Alex, and Vinita Damodaran. "Introduction: Human-nature Interactions through a Multispecies Lens". *Conservation and Society*, vol. 14, no. 4, 2016, pp. 293–304.

Barrett, Louise. *Beyond the Brain: How Body and Environment Shape Animal and Human Minds*. Princeton UP, 2011.

Besnier, Niko. "Language and Affect". *Annual Review of Anthropology*, vol. 19, no. 1, 1990, pp. 419–51.

Chattopadhyay, Swati. "Visualizing the Body Politic". *Making Place: Space and Embodiment in the City*, edited by Arijit Sen and Lisa Silverman, Indiana UP, 2014, pp. 44–68.

Chomsky, Noam. *Aspects of the Theory of Syntax.* Cambridge: The MIT P, 1965.

Diffloth, Gerard. "Notes on Expressive Meaning". *Chicago Linguistic Society*, vol. 8, no. 44, 1972, pp. 440–47.

Dingemanse, Mark. "Redrawing the Margins of Language: Lessons from Research on Ideophones". *Glossa: A Journal of General Linguistics*, vol. 3, no. 1, 2018, pp. 1–30.

Haraway, Donna. *When Species Meet.* U of Minnesota P, 2008.

Hirose, Naoya. "An Ecological Approach to Embodiment and Cognition". *Cognitive Systems Research*, vol. 3, no. 3, 2002, pp. 289–99.

Jaaware, Aniket. *Practicing Caste: On Touching and Not Touching*. Fordham UP, 2019.

Jameson, Frederick. *The Antinomies of Realism*. Verso Books, 2013.

Jakobson, Roman. "Metalanguage as a Linguistic Problem". *The Framework of Language.* Michigan Studies in the Humanities, 1980, pp. 81–92.

Kataoka, Kuniyoshi, Keiko Ikeda, and Niko Besnier. "Decentering and Recentering Communicative Competence". *Language & Communication*, vol. 33, no. 4, 2013, pp. 345–50.

Kohn, Eduardo. *How Forests Think: Toward an Anthropology Beyond the Human*. U of California P, 2013.

Küpers, Wendelin. "Embodied, Relational Practices of Human and Non-Human in a Material, Social, and Cultural Nexus of Organizations". *Culture: The Open Journal for the Study of Culture*, vol. 2, no. 2, 2016. geb.uni-giessen.de/geb/volltexte/2016/12352. Accessed 24 May 2020.

Lear, Johnathan. *Radical Hope: Ethics in the Face of Cultural Devastation*. Harvard UP, 2006.

Low, Setha. "Placemaking and Embodied Space". *Making Place: Space and Embodiment in the City*, edited by Arijit Sen and Lisa Silverman, Indiana UP, 2014, pp. 19–43.

Massumi, Brian. "The Autonomy of Affect". *Cultural Critique*, vol. 31, 1995, pp. 83–109.

Menary, Richard, editor. *The Extended Mind*. MIT P, 2010.

Morris, Brian. *Animals and Ancestors: An Ethnography*. Berg, 2000.

Sanders, John T. "Affordances: An Ecological Approach to First Philosophy". *Perspectives on Embodiment: The Intersections of Nature and Culture*, edited by Gail Weiss and Honi Fern Haber, Routledge, 1999, pp. 121–42.

Streeck, Jurgen. "Depicting Gestures: Examples of the Analysis of Embodied Communication in the Arts of the West". *Gesture*, vol. 9, no. 1, 2009, pp. 1–34.

Tsing, Anna. *The Mushroom at the End of the World: On the Possibility of Life in Capitalist Ruins*. Princeton UP, 2015.

Windsor, W. Luke. "An Ecological Approach to Semiotics". *Journal for the Theory of Social Behaviour*, vol. 34, no. 2, 2004, pp. 179–98.

Van Dooren, Thom. "Care: Living Lexicon for the Environmental Humanities". *Environmental Humanities*, vol. 5, no. 1, 2014, pp. 291–94.

———. "A Day with Crows—Rarity, Nativity and the Violent-Care of Conservation". *Animal Studies Journal*, vol. 4, no. 2, 2015, pp. 1–28.

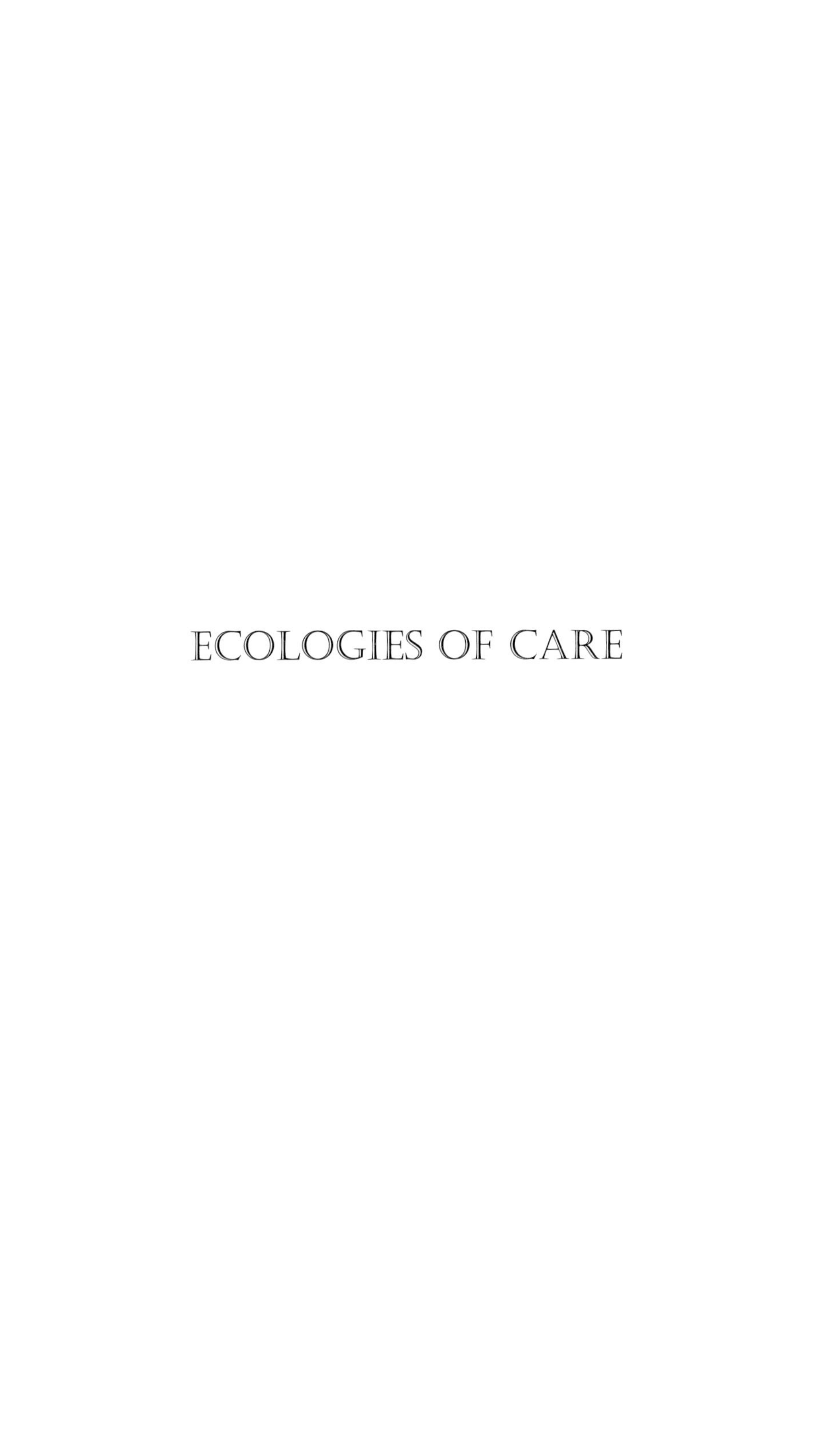

ECOLOGIES OF CARE

ONE

ETHOS, PATHOS, LOGOS

AFFECTIVE AND EMOTIVE ETHNOGRAPHIES OF HUMAN–MACAQUE LIFEWORLDS

Ishika Ramakrishna, Ajith Kumar, Maan Barua & Anindya Sinha

'There will always be some resentment', said Sanjay, a middle-aged farmer, one afternoon as one of us surveyed the damage the macaques had inflicted on his coconut plantation. His voice was calm. His body, initially tense while he calculated the losses faced due to the monkeys that frequented his land, was now relaxed. However, the statement—apparently simple—came from a place of repeated personal experience, cultural influences, familial history and an understanding of the macaques' needs. In one brief sentence, Sanjay had assessed the costs and benefits of a shared living space with macaques and adequately summed up how he felt.

The farmer and the Nicobar long-tailed macaque (*Macaca fascicularis umbrosus*), the subject of this discussion, both inhabit the island of Great Nicobar in the Andaman and Nicobar Islands of India. The accounts of several islanders like Sanjay made it evident to us that people's perceptions towards nonhuman primates—their emotions shaped by direct interactions, actual initiated behaviours towards the macaques, and their emergent affectual geographies—combine, often in strange and complex ways, to generate a unique individuality in every human–nonhuman relationship that develops.

EMBRACING VARIABILITY IN A RELATIONSHIP

Multispecies coexistence often demands that humans assign beliefs, emotions and affective responses to other species in order to predict their future actions, especially when they lack direct access to the thoughts of others (Ghuman et al.). This becomes a necessity, especially in cases where non/humans find themselves in close proximity to one another in the everyday. Such an ability to construct minds, whether in oneself or in others—or what has been scientifically termed as a capacity to have a 'theory of mind'—adds further complexity to how real-world human–nonhuman relationships develop over time (Premack and Woodruff; Ghuman et al.). This also lends itself to a certain level of subjectivity in people's assessment of nonhuman emotions, internal states or causal drivers of behaviour, all shaped by their own personal biases.

This subjectivity could be further enhanced by a rather human tendency to identify with the nonhuman—to consider the other as being similar to oneself—or, in other words, to anthropomorphise wherein human-like qualities are attributed to nonhuman entities. Anthropomorphic tendencies appear to be highly amplified in human–nonhuman primate relationships owing possibly to their shared evolutionary histories, which may have in turn, led to apparent similarities in their morphological and behavioural features (Madden). Nonhuman primates (hereafter primates) thus often become literal and figurative kin to people, taking on important positions in the religious and cultural beliefs of human communities that share space and live with other primate species (Fuentes, "Ethnoprimatology"), although they may occasionally have overlapping and often conflicting interests with competitive interactions over food, space or other resources (Aureli and Smucny). It was however, immediately apparent to us during this study that anthropomorphism was not an invariable attribute, but a dynamic, emergent projection continually crafted in practice. It was also possible and, we believe, important to read this anthropomorphism in affective terms.

The spectrum of interactions between people and primates is extremely diverse and yet, there seems to be an overt focus on 'conflict' in defining these human–nonhuman relationships (see also Banerjee and Sinha in this volume). Predefining interactions in this manner biases our thinking, making it challenging to understand situational nuances that appear to be the norm (Gusset et al.). Furthermore, individual humans and primates show great diversity in their own behavioural responses to the other, principally stemming from their own experiences and personal adaptability and shaped by their intellectual capacities. Amidst rising levels of direct non/human interactions; rapid anthropogenic development and loss of natural habitats; and in the case of humans, the cultural significance of primates to particular communities, this diversity is further pronounced.

Both macaques and langurs, commonly distributed across the Indian subcontinent, are worshipped by Hindus and hold immense cultural significance—being revered due to the symbolism associated with the monkey god, Hanuman. This further complicates the nature of the interactions that occur between them, in turn influencing the future course of their relationships (Saraswat et al.). What often remains unrecognised, nevertheless, is the manner in which these relationships reflect an amalgamation of mutual tolerance, compassion, fear and occasionally, even an upsurge of violence (Gumert et al.).

India is home to approximately fourteen species of diurnal primates—largely macaques and langurs—that have had a long history of interactions with people owing to their close proximity to one another and their mutually high population densities across the country (Southwick and Siddiqi). Wildlife populations, like those of the bonnet macaque (*Macaca radiata*), rhesus macaque (*Macaca mulatta*) or the common or grey langur (*Semnopithecus* species), which have become increasingly urbanised over the past decades (Srinivasaiah et al.), have attracted much attention for their negative interactions with people across the subcontinent (Sinha and Mukhopadhyay; Saraswat et al.; Radhakrishna). Such negative perceptions have largely emerged due to the increasing practice of crop foraging by primate groups; their unwanted entry into gardens or kitchens in search of food; or the display of aggression by habituated troops in religious or tourist sites, resulting in almost constant clashes between people and primates (Hill, "Primate Conservation"; "Primate Crop Feeding"). While such negative interactions are generally encountered across the country, what remains largely ignored are the nuances of the prevalent relationships that are typical of particular human communities and primate species groups, which have coexisted over relatively long periods of time, further contextualised by specific local ecological regimes that often change over shorter time scales (Schilaci et al.; Fuentes, "Ethnoprimatology"; Baker et al.; Radhakrishna; Sinha et al.).

The subjectivity and individuality that contribute to the variation found in these interactive systems indicate the need to move away from approaches that have traditionally typified non/human interactions. Thus, this chapter celebrates this variation, as it is not only reflective of the remarkable behavioural plasticity and ecological adaptability of non/human primates that have shared their lives over centuries, but can also allow us to understand, interpret and contextualise human–primate interactions in the light of their historical, socioeconomic, ecological, cultural and temperamental influences.

EMOTIONAL AND AFFECTIVE ETHNOGRAPHIES

Interactions between people and nonhuman species often become emotionally charged. It has, in fact, been argued that the emotional relationships that people have with their environments shape their own societies by helping them *know* and *do* (Pile 6, emphasis added). This allows for a mechanistic approach to understanding people's behaviours and their underlying empathy, which contributes to how real-world interactions emerge in a shared landscape. It is also geographical in spirit, whether emotional, affective or socioecological. Such an approach, we believe, is of great heuristic value as the human condition and the worlds within which people reside are built up of their emotions; these could include love, grief, happiness, fear, anger and more complex admixtures of such feelings. Emotions, thus, potentiate behaviours and actions (de Waal; Kremer et al.).

The French poststructuralist thinker, Gilles Deleuze, whose work on affect has resonated across the social sciences, makes a number of important connections between affect and ethology. He argues, for example, that 'studies ... which define bodies, animals, or humans by the affects they are capable of, founded what is today called *ethology*. The approach is no less valid for us, for human beings, than for animals, because no one knows ahead of time the affects one is capable of' (125). Affect, when understood in ethological terms, also implies bidirectionality. Animals can be thought of as experts at reading bodies, whether their own or those of other sentient beings but, importantly, they can also learn to be affected by their human counterparts (Despret). We thus examine ethology here as a study of affect—a means through which animals lead sentient lives—and through a comprehension of which we may be able to understand how human and other-than-human worlds are co-composed; where the animal is taken to be knowledgeable and even capable of producing outcomes (Barua and Sinha), although abilities to do so differ. Even though we might not entirely know what animals think and feel, an inquiry into affect may enable us to examine the 'everchanging processes human and nonhuman bodies undergo as they experience, encounter, and perform life among other bodies within material space' (O'Grady 1).

Affect is an 'intensity' (Massumi)—a sentient expression that is almost liberated from systems of representation. Affects swirl between and through bodies—both human and nonhuman—transcending the division between oneself and external interacting bodies: an ebb and flow between an agent and its environment (Solomon). Affects are performed and felt and can be invented or engineered as they course through the rhythms of everyday life and experience. They can thus be distinguished from emotions in that they are non- or pre-cognitive in nature whereas emotions, as well as the treatment of affect within cognitivist literatures, can be considered objects of the mind (Pile). It is the former understanding of affect, as a transpersonal (Anderson) and trans-species capacity (Despret), that we refer to in this chapter.

Emotions, in contrast, are expressed feelings that are both conscious and bodily experienced (Pile). It has also been considered a multicomponent—subjective, physiological, behavioural and cognitive—response to a stimulus or event that is typically of importance to the individual, often valenced as pleasant or unpleasant and variable in activation/arousal and duration/persistence (Paul and Mendl; Kremer et al.). Emotion is clearly more tangible and communicable across individuals of the same or different species (Kremer et al; Sinha et al.). This is of particular interest in the case of human–nonhuman primate interactions, as the flow and transfer of these states across species are likely to be (1) more compatible, though nuanced, owing to the evolutionary, physical and interpretative similarities between these species; and (2) easily observable in both people and nonhuman primates.

In this chapter, we argue for (1) the prime importance of conducting affectual and emotional ethnographies of human beings immersed in a multispecies existence, within which they interact with other-than-humans and co-construct their lifeworlds with them; and (2) the crucial necessity of invoking and documenting individuality in human–nonhuman primate relationships. We exemplify these endeavours with a case study of the complex interactions presented by diverse human communities on the island of Great Nicobar, part of the Andaman and Nicobar Islands, off the south-eastern coast of India, and the endemic Nicobar long-tailed macaque (*Macaca fascicularis umbrosus*) that lives amidst them. As part of a larger study that examined the nature and extent of these human–macaque interactions, we conducted open-ended interviews with 80 individuals—44 women and 36 men ranging in age from 21 to 72 years—residing across the 9 villages of the island. These interviews covered topics spanning people's histories of interactions with the macaques; their vernacular knowledge of macaque ecology and behaviour; their present-day interactions with them; documentation of folklore and tales relevant to these interactions; and any mitigation measures implemented to aid in the alleviation of losses sustained due to perceived conflict with the macaques.

HUMANS AND MACAQUES ON THE GREAT NICOBAR ISLAND

In countries like India, as previously established, langurs and macaques have historical, cultural and religious significance across human communities (Gumert; Saraswat et al.). While in many cases the relationship between the species is commensal (Fuentes, "Human-Nonhuman Primate"; Gumert et al.), they can take on many other forms and derive from a broad spectrum of interactions ranging from the mutualistic to the parasitic. Our current understanding of these relationships, unfortunately, revolves around the outcomes of studies that have principally recorded events of nonhuman primate crop-foraging and their resultant losses (Hill, "Primate Conservation"; "Primate Crop Feeding"). This focus on crop-foraging (or crop-*raiding*) behaviour and on people's perceptions of primates as 'vermin' or 'pests' does not, however, entirely speak of the nuances of how these species are understood or viewed by the local communities (Hill and Webber). Moreover, human perceptions of other primate species tend to depend on a multitude of factors, including the history of the focal area, the socioeconomic status of the people concerned, their religious and cultural inclinations, past experiences with wildlife or the concerned species, and the nature and extent of the current interactions among them (Arroyo-Rodriguez and Dias).

The Nicobar long-tailed macaque is an endemic subspecies found on the Katchal, Little Nicobar and Great Nicobar islands. These three islands are home to ~5,000 individuals of this primate with most of their subpopulations residing within human inhabited areas (Umapathy et al.).

On the largest of these islands, Great Nicobar, the macaque now occurs amidst several human communities along the south-eastern strip of the island. Several linguistically, culturally and socioeconomically distinct communities from the Indian mainland—Punjabi, Bengali, Tamil, Malayali, Bihari, Telugu and Maharashtrian—have occupied land and houses (segregated in status) across the nine villages of the island since the 1970s, apart from settlements of the Indigenous Nicobarese and Shompen people. The social divide between the

Indigenous and settler communities, however, has become less apparent in the years that followed the destructive tsunami of December 2004—a natural calamity that appears to have also majorly shaped the evolution of interactions between the local people and the macaques on the island (Mishra et al.).

The human settler communities and the macaques were both adversely affected by the tsunami of 2004, suffering losses of resources and habitats (Velankar et al.). The people were forced further inland while rebuilding their living spaces, encroaching into forested areas in the process. The macaques' staple *Pandanus* fruits—which occurred abundantly all along the coastal areas—were destroyed as well, leaving behind very few surviving individual trees (Porwal et al.). While there was a certain degree of human–macaque interactions even before 2004—usually manifest as occasional crop foraging by the macaques (Umapathy et al.)—the post–tsunami settlement patterns significantly enhanced their frequency and intensity. Moreover, two comparative surveys conducted across the three islands in 2006 and 2014 showed that the encounter rate of macaque troops increased from 0.23 troop/km (recorded in an earlier survey in 2003) to 0.30 troop/km subsequently (Umapathy et al.; Velankar et al.). This, we believe, could have triggered the increasing animosity expressed by the local human communities towards the macaques coupled with a reduction in tolerance of their constant presence, which we discuss later.

Studies from other parts of mainland India and regions of Southeast Asia describe cases of high-intensity 'conflict' between primate species like the rhesus macaque, bonnet macaque or other subspecies of the long-tailed macaque (Marchal and Hill; Singh and Thakur; Saraswat et al.; Singh). In most of these cases, such conflict has led to significant levels of financial losses, negatively affecting people's livelihoods. This has also been accompanied by a shift in their attitudes towards monkeys, from one of positivity to that of often intense antagonism (Richard et al; Manral et al.). While the current situation on the island of Great Nicobar is far from dire, it shares key milestones in the development and manifestation of negative interactions between people and primates from other parts of the world. Our study, therefore, hoped to provide fine-grained descriptions of human–primate relationships in order to understand the interplay of the multitude of factors affecting interactions between people and the Nicobar long-tailed macaque.

Following the tsunami, the social hierarchy of the human communities—which had been painstakingly established since the 1970s when people first began migrating from the mainland—was broken down. All the communities now found themselves living together under similar conditions and with comparable resources at their disposal, resulting in the beginning of novel, shared cultures and changes in the original social dynamic that persists even today. We were thus able to identify how people's changing circumstances, emotional and affective states and religious affiliations influenced the manner in which they responded to interactions with macaques and to the environment they lived in.

In the final section of the chapter, we choose to interpret the variability in people's emotive and affective states in response to their interactions with macaques, in terms of Aristotle's three methods of persuasion—*ethos*, *pathos* and *logos*. Whilst the personal arguments constructed internally by individuals, often unconsciously and verbally expressed on occasion, could represent a spectrum varying from logical reasoning to pure sentiment with shades in between, we analyse them in retrospect by examining the motivations behind why people *feel* and *behave* in certain ways that they do. Understanding these cognition-based emotive and pre- or non-cognitive affective drivers behind people's perceptions, beliefs or actions towards macaques could also potentially help, we believe, in determining context-specific mitigation measures to minimise negative interspecies interactions in the future.

Experience: The History of a Relationship

The Great Nicobar Island is home to two Indigenous communities, the Nicobarese and the Shompen (Elanchezhian et al.). In April 1969, 23 families of ex-army personnel from Punjab were offered 11 to 14 acres of agricultural land each on the island—the primary motivation behind this being the perceived necessity to have Indian occupants settled in this border area. The island's first settlement was formally established in 1971 (Saini). Subsequently, a steady inflow of people from the mainland and the Andaman Islands began to occupy the island, either as government employees, farmers and plantation owners, or as labourers to aid with development. Over the next three decades, the south-eastern strip of the island that ran along the coastline was developed extensively to accommodate all those settled and their needs, with most of the initial settlers engaged in agricultural activities alone. They maintained paddy fields, coconut (*Cocos nucifera*) and areca nut (*Areca catechu*) plantations, and developed mango (*Mangifera indica*), guava (*Psidium guajava*) and chikoo/sapota (*Manilkara zapota*) orchards.

Historically, from 1970 to 2000, the Nicobar long-tailed macaques were reported to only rarely venture out of the neighbouring forested areas. The macaques would enter paddy fields, plantations or orchards usually in the absence of humans. There was mutual fear between the two species. Most agrarian families kept domestic dogs that would chase monkeys, occasionally killing them.

The macaques were not typically fed by people on the island—with the exception of a few families and priests who lived in the main Campbell Bay area near some temples. The first sign of losses due to crop foraging, however, appeared when macaques began to venture into unattended paddy fields—which did not expect such depredation—to feed on rice grains. Precise quantification of these losses had never been attempted and our respondents were unable to recollect the monetary or other losses that they had sustained. With the exception of a minor 6 per cent of the households, people felt that the extent of crop damage—in both paddy fields and vegetable farms—was minimal during the initial stages of their settlement.

On 26 December 2004, a massive earthquake caused a tsunami that was felt all over the island of Great Nicobar. Many families lost several acres of land or were left with uncultivable land due to waterlogging (Malik et al.). It took seven to nine years of redevelopment and relief provisioning to resettle the surviving local people into tsunami shelters; a large number of people began to live in camps in the Campbell Bay area, abandoning their fields and weathered homes. These shelters were clustered into nine villages, as opposed to the more spread out distribution of homes laid within farmland, prior to the tsunami. The macaques were further disadvantaged by the extensive forests being 'torn', 'broken', 'cleaned' and 'destroyed' to make room for more human habitations above sea level.

With the rapid establishment of new human settlements on the island, people began to notice an increase in macaque populations, along with the frequency of their visits. Moreover, while the monkeys had tended to fear people earlier, they no longer kept a cautious distance from human settlements. There were also changes in their foraging strategies that attracted attention. They had learnt to process coconuts and 'steal' chicken eggs from coops and had increased their consumption of fruits and vegetables. Over time, the macaques increased their extent and duration of foraging within coconut plantations, occasionally visiting 60 to 80 per cent of the palms, as opposed to feeding on no more than 10 per cent of the trees in the past. Most respondents believed that there were fewer food trees and fruits available to the macaques in the forest, due to which they had learnt to forage in farms and plantations. Some claimed that the macaques had gradually learnt to consume, process and 'enjoy' human-origin foods, which they preferred to the resources available in the forest.

In general, a large majority of our respondents had come to believe that the macaques were currently consuming significantly more human-origin foods, as opposed to 20 to 30 years ago, primarily due to the gradual reduction in forest cover where they once occurred—the macaques thus had no place to go to and were forced to live with people. And, as could be expected, they had to now take recourse to 'scaring' and 'threatening' people and to committing 'crimes' that had begun to seriously affect the local populace.

The people's perspectives on macaques, we found, could be broadly categorised into positive, negative or neutral, although they were far more nuanced than these labels convey. More importantly, they were based on a wide array of their experiences and knowledge of the species—from their distinctive appearance to their intriguing behaviour.

As these macaques were dark grey to black in colour, some respondents had derogatory connotations to offer while referring to their general appearance and, thereby, implied intentions. They were often considered to be 'greedy' or 'lazy', as they repeatedly visited areas where they could easily find food, such as unattended gardens or coconut trees. Certain experiences that the people claimed to have had or heard about, including macaques pursuing women with flowing clothing, pulling on their dress or 'harassing' them by bared teeth displays, led them to believe that they could be 'perverts'. People also thought of them as 'opportunistic' creatures who utilised every possibility to take food from people, just as 'thieves' would. Interestingly, they were also thought of being 'rebellious' as they could clearly learn fast and use their new-found knowledge to upturn human order and disrupt well-established practices in order to create an environment that was to their advantage.

A recurring theme throughout the interviews was that of 'pity'—in a way reflective of a sense of empathy—towards the monkeys, leading several families to tolerate their foraging in gardens as they clearly needed the food to survive. Three male respondents recounted tales of other humans stealing coconuts from their collection piles, making them question how people were any different from macaques, with respect to the 'crimes they commit'. Anthropomorphism surfaced in several interviews, with up to 41 per cent of the respondents drawing evolutionary, behavioural and physical parallels between human beings and the macaques. Several people even thought of them as being 'smart', 'intelligent' or 'creative' beings, who could carry out nearly any task that a human being could.

The negativity expressed by certain respondents primarily concerned a resentment against *their* land being occupied by the macaques, thus depriving the locals of the ability to lead 'normal lives', reminiscent of a sense of longing and personal geography. Some felt that living in such close proximity to macaques or other wild animals was highly 'unnatural' and in urgent need of amendment. Some of the most recurrent words, evocative of the complex, largely negative emotions experienced and used to describe the macaques' actions included 'crime', 'destruction', 'ruination', 'dance', 'uprooting', 'chaos', 'thievery', 'hiding', 'sneaking' and 'messing up'.

Finally, there was a distinct sense of disappointment palpably expressed by many respondents and which manifested occasionally as anger, fear or emotional distance from the government-run Department of Environment and Forests in Campbell Bay. People visualised the Forest Department as being a dominating, authoritative body far removed and occasionally alien to them, with individual department members—with whom attempts were made to establish rapport—being transferred out of the island before any solutions could be put in place.

ETHOS: PILLARS OF WISDOM

Human social attitudes and behaviour are coded but they are also based on written and tacit rules. The ethos of people's arguments for how they perceive macaques are thus often grounded in the authoritative pillars that have been laid down and acknowledged at various scales—from the individually determined or personal to the community level—including culture, religion or government laws. Ethos, however, could also be read as the cultivation of habits and behaviours (Guattari). We were thus able to discover unique ways in which people understood and appreciated macaques, emerging from interactions with one another at the individual level and manifesting, in this particular situation, when our respondents reported on (1) their knowledge of the species' ecology and behaviour; (2) direct observations of species behaviour; and (3) their past experiences or historicity in the spaces shared with macaques.

Vasanti, a fisherwoman and a mother in her thirties, said, 'These really large monkeys would come right in front of me sometimes, as though saying, "If you cause harm to us, we're going to bite you badly". But now I've started praying to Hanuman so much that they could never harm me. Ever since I've been devoted to Hanuman, nothing untoward has happened to me and I'm sure nothing bad will happen in the future either.' Her beliefs and insights are of interest here, as she not only associated the macaques with Hanuman but also extended her faith in the monkey god to keep her safe from his own kind. Furthermore, she had given up on any mitigation measures to keep the macaques away from her kitchen garden and instead, sought refuge in prayer to keep herself and her children safe. Another respondent, Jai, a male farmer, reported, 'You should see the monkeys eating coconuts! They drink coconut water so fantastically and then throw them down! The other day, one of the coconuts was about to fall on my head, I just about escaped. God saved me that day. I was standing there [pointing into his field] and the coconut fell, just missing me.' Jai's words show not only how a potentially negative experience was ameliorated by a sense of wonder and gratitude owing to his religious faith, but also the affects that an encounter with macaques sparked.

Religion appeared to serve as a moral compass by which people determined whether their actions, directed at the monkeys, were justified or could be frowned upon. It must be remembered however, that religion does not always specify predetermined categories of belief but is reiterated through repeated performance and practice. It could, nevertheless, work as a filter of empathy in the real world—which it did on the island—allowing people to draw from the cultural or religious importance of the macaques and relate those beliefs to the events unfolding in their backyards. The population on Great Nicobar is predominantly Hindu (69.45 per cent), followed by Christian (21.28 per cent), Muslim (8.52 per cent) and Sikh (0.34 per cent) as per the 2011 Census of India. The island houses several temples—of which three are dedicated to the monkey god Hanuman—one gurudwara, one mosque and three churches. The predominance of Hinduism and Christianity on the island was evident from the observation that the festivities at the local temples and churches were the primary social events for people of all religions. This seems to have instilled a secular collective culture on the island, whereby every community was both sensitive and aware of the other religious affiliations on the island.

The presence of primates in Hindu mythology, with people—both Hindu and non-Hindu—having been brought up on tales of Hanuman, often strongly influences the outlook that many community members have towards macaques (Chauhan and Pirta, "Agnostic Interactions"). Such beliefs additionally determine how extreme a household's mitigation measures against the macaques could potentially be (Chauhan and Pirta, "Socio-Ecology of Two Species"). A similar comparison of the local long-tailed macaques to Hanuman was a common feature in several of our interviews. Some individuals thus staunchly believed that all monkeys were Hanuman incarnate and that they should be treated with due respect and protected, irrespective of the losses they caused. Others, while being reverential to the godly macaques, were not averse to mitigating their losses by pelting an occasional stone or aiming a catapult at a macaque, although they would never advocate the killing of monkeys. These accounts once again reiterated our belief that religious scriptures may not often directly dictate people's actions. On the contrary, such scriptures are practically rewritten by the performances drawn out of people by the prevailing conditions. For instance, one of our female reporters, Surekha, whose family subsisted on agriculture, reflected, 'No one has ever killed a monkey here. This whole place, Shastri Nagar, is a Hindu area. Hindu people don't kill monkeys. If we are angry, we just chase and throw stones at them at the most.' On certain occasions, religious sentiments were employed as per convenience or conditionally while in a small subset of interviews, people turned to their family gods to keep themselves and their assets safe rather than directly viewing the macaques as sacred.

A comparable number of respondents, however, logically concluded that the Nicobar long-tailed macaques in particular, could not be manifestations of Hanuman as they clearly did not possess the 'pious' qualities that a god was meant to have. These macaques were despicably 'non-vegetarian', as they consumed eggs, young chickens, fish or crabs. They were also vastly different in their appearance from the stately mythological depictions of the monkey god, leading individuals to extend little allowance for the losses they suffered at the hands of the macaques. In other words, there were contradictions between the usual anthropomorphic projections and what macaques did in actual practice.

Hints of righteous environmentalism could be seen in about 24 per cent of the respondents, who showed explicit awareness of how people had manipulated natural spaces through developmental activities. Their outlook towards macaques and their plight was reflected in their responses to the mitigation measures adopted. George, a person working with the government Electricity Department in the main town of Great Nicobar, said, 'The monkeys have been here from before we arrived. They were here long before man, you may have read about this in books as well. People say that the monkeys are Hanuman, which is absolutely right—we won't kill them.' At times, this awareness was combined with people's knowledge of macaque ecology and needs, obtained through regular observations. To this effect, Alka, a female nurse from the island's medical centre, said, 'We made our house in the jungle. Where will the monkeys go? This is their place too. They will keep on coming here. They eat eggs, vegetables, young chicks. People used to say earlier that this monkey is like god, it will not eat anything here. But what will they do if they are hungry?' In this interview, the nurse further rationalised that while she had always revered Hanuman and believed that all life was sacred, she would not consider the macaques to be manifestations of god. To her, they were inconvenient wildlife—years of experience had seasoned her religious outlook towards these primates. Shades of empathetic environmentalism possibly also led certain individuals to feed macaques—even though this encouraged them to repeatedly visit their homes and farms—or simply chase the macaques away, although they regularly caused significant financial losses to these families.

Considerations of legality constituted, perhaps, the most straightforward of reasons why people chose not to harm the macaques despite feelings of resentment, frustration or annoyance towards them. Adherence to the law, which deemed it illegal to harm/kill/capture individuals of this species—protected by the Wildlife Protection Act of India, 1972—came across as purely obligatory rather than as an internalised fact that the people morally agreed with. By extension, several people felt that the Forest Department 'owned' the monkeys and it was their 'duty' to protect them. A male islander, Vijay's thoughts on this matter succinctly demonstrated the predominant reasoning behind why people followed the law:

> What would anyone do after catching the monkeys? The forest people will come after them. If a dog kills a monkey and the forest people hear about it, they immediately come with the police. Raja's dog killed a monkey and a fine of 20000 was imposed on him. We should be compensated then for all the damage we incur because of the monkeys. Instead, the

> Forest Department will come immediately to impose fines. Why don't they tie up their monkeys? But they won't do that.

Beyond the legal and the religious, certain individuals took recourse to their lived experiences to establish baseline expectations of their interactions with macaques. Their detailed observations—noted out of necessity, habit, wonder, joy, or simply an overlap in space and constituting local ethologies of the macaques—contributed to a particular *ethos* (Guattari) and provided them with underlying motivations that guided their feelings, perceptions or actions. We suggest that these individual variations in lived experience led to the formation of unique affective states in people, predisposing them to certain kinds of positive or negative behaviours.

Time was another factor that played into perceptions and pre-mediated behaviours directed towards the monkeys. For those who have resided on the island for several decades, early memories of interactions and the gradual acknowledgement of how the landscape has evolved appeared to determine both their opinions and actions. Susheela, a middle-aged mother of three and wife of a coconut plantation owner, firmly stated, 'Living in conjunction with the monkeys was the norm before the tsunami, which is why we'd never even utter the idea of killing them. It may have changed for a lot of people now, but that's not how I remember things being.' Similarly, Noel, a Christian man, who claimed to be fascinated by macaque behaviour, begrudgingly said,

> Earlier, nearly every day or once in two days, I'd feed the monkeys near the open graveyard. But these days if I were to feed them here inside the village, they'd develop the habit of coming here. Instead, we all chase them away. But then again, it's not like they run away; they'll just sit at a slight distance and stay put. Knowing that we don't catch them, they don't stray too far or run either.

PATHOS: VOICES OF PASSION

While the ethos basically provided a baseline framework upon which people's perceptions, emotions and affective states resided, the pathos underlying their individual arguments was often stronger. The role of pure sentiment, drawn from personal experience, was arguably the most significant driver of how and why people formed specific perceptions or opinions of the macaques.

During the course of inductive coding for the qualitative analysis of people's accounts, we found an incredible array of emotive states being expressed including wonder, horror, desire, amusement, defeat, fear, sympathy, admiration, exclamation, exasperation, anthropomorphism, curiosity, disappointment, sadness, anger, indignation, denial, anxiety, exaggeration, expectation, amazement, frustration, romanticism, aspiration, indifference, annoyance, neutrality, power, fascination, judgement, ambition, disbelief, joy, compassion and care.

Malati, a young woman, who had moved to the island of Great Nicobar only two years prior to our conversation, exclaimed, 'I really wish sometimes to just catch one baby monkey and take care of it at home! I think it would be wonderful to have a little monkey climbing over things in the house! It would probably always sit on my shoulder or hold onto me all day.' A maternal tendency towards the species was also recorded in a few female villagers, who had either heard of cases where infants were 'adopted' by people or envisioned a possibility in which they could care for one. Feelings of compassion emerged indirectly through observations made by women, such as Sita, 'Small monkeys that cling to their mothers are also seen. Sometimes if the young monkeys are left behind, the group will come back for them. I have seen this in my old house. Four to five dogs went after a young monkey but the whole group came and saved it. It felt good to see that.'

Kanthu, a young policeman, was enamoured with the monkeys' skills, being part of the approximate 40 per cent of the people who anthropomorphised the species. He said, the monkey enters the kitchen, with expertise opens the fridge door and takes away some eggs. The monkeys do all this fantastically! Even our brains won't work as well as this. Their minds are very sharp.' Another statement, made by a neighbour of Kanthu's, also belonged to a series of incredulous descriptions of the macaques' behaviour hinting at some disbelief in their proclaimed similarities to human mannerisms. He said, 'The monkeys tear the skin of the coconut and eat it despite the skin being so hard. How they are able to do so, I don't know. Their hands do not pain. They also throw the coconut to break it open and then eat it. They drink water from the coconut just like a human being.' While this description revealed a sense of wonder and fascination at a proximal level, it also harboured underlying feelings of awe and fear for what the macaques appeared to be capable of.

Anthropomorphism, however, could be threatening for the macaques when it involved certain human sentiments and expectations.

While the words of Sita, Kanthu or Malati largely revealed positive perceptions of the monkeys, relating to the macaques closely could also result in having human-like expectations of them—those that would obviously not be met. When the macaques failed to live up to such human conduct and courtesy, people's opinions of them fell lower still. Preeti, a newly married woman in her twenties, claimed, 'They cause damage. They make us cry. They won't leave anything for us to eat.' Her words rang of utter disappointment at the agency she had assumed the macaques had. Having also imparted intentionality to the monkeys (Sinha et al.), Preeti's sense of loss was not just for the material but, more importantly, it was emotional: wrought by the macaques having failed to meet her expectations.

Beyond anthropomorphism and the imposition of human expectations on these nonhuman beings, people also seemed to preferentially hold on to the memories of negative interactions over those more neutral or positive. These tended to create a negatively biased foundation on which their imagined future interactions rested. Bhupender, a Punjabi farmer, who had been cultivating coconuts for two decades on the island, warned us,

> The monkeys are quite dangerous. You should never venture into a group of monkeys. If even one starts calling, the whole gang will come forward. I get extremely scared. Once a big group chased me for a short distance! I get even more scared thinking of all the things that they could have done to me, like scratch me, tear at my skin or bite me. And now I simply run away in such situations!

The documentation of the unusual combinations of emotions that led to each individual's unique perceptions of the macaques made us wonder how such deeply positive and negative feelings could seamlessly coexist in the daily lives of the islanders. While it would appear contradictory to feel empathetic towards the monkeys and yet pelt stones at them in anger, such interactions did seem to be in harmony if we considered a real-time rationalisation of one's personal principles, experiences and emotions, driven simultaneously by a sense of identification with the macaques on the one hand and by feelings of being deprived of one's precious resources on the other.

LOGOS: ARGUMENTS OF REASON

A final angle that explained how people's emotive states could translate into action was that of rational thought driven by logical reasoning.

While a rationalisation of one's beliefs and actions appeared to bridge the ethos (the foundations of such beliefs) and pathos (the ways in which people justified their actions), this transition was also reinforced by aesthesis—the innate capacity of individuals *to* affect and *be* affected. The lived experiences of the people combined with their perceptions, emotions or temperament and guided by their sociocultural and religious principles, often manifested themselves in certain attitudinal positions and on-ground action. It thus appeared, in these instances, that the final decision of how one should actually behave towards the macaques was taken after a careful consideration all the different influencing factors—whether consciously or subconsciously—in an apparently logical and rational manner.

Suman, an areca nut plantation owner, thus said, 'In my experience, I've learned that we can walk through any number of monkeys, as long as we don't look at them. I've noticed that if you look around you constantly, make eye contact with them or stare at them, they start reacting to you. If we simply walk straight through them, they won't look at us or take notice either.' Suman's learnings were drawn from a spirit of curiosity and experimentation that was particularly common amongst the farmers who were interviewed. Certain forms of hypothesis testing, we argue, were also evident in the way people attempted variations of interaction with the macaques in order to arrive at the most effective manner of behaving towards them. Tacit and coded in practice, it appeared to us that individuals like Suman were unconsciously learning to be affected by the macaques, developing the capacity to read macaque bodies and let himself be read by them—Guattari's ethico-aesthetics in action.

Apart from being able to rationalise their own behaviours towards the macaques, several people also provided logical explanations for why macaques behaved in certain ways, thereby dissociating anthropomorphic expectations from the actual actions that they had observed—fine examples, once again, of vernacular ethologies. An account from Philmon, a seventy-two-year-old fisherwoman, for example, went thus, 'Once the monkeys had got into the neighbours' house and damaged everything! No one was at home. When they do things like this, it's obvious that people will feel anger towards the monkeys. But that's also their fault, right? If you leave things open and leave the house empty, something like this is bound to happen.'

Hermon, an aged man from the same fishing village, recounted,

> An old lady used to go out carrying her food items on her head, and the monkeys would just come and eat up all the food. They didn't do anything to the old lady though. They never really harm anyone. Since they're hungry, at times they'll scare or threaten people, but they don't mean any harm. If they wanted to bite people, can you imagine how many people would've been bitten until now?

Through these conversations, we gathered that much of the *logos* of people's arguments tended to exonerate the macaques, the Forest Department, as well as various deities from blame for negative interactions between people and macaques. The people themselves were held accountable for the roles they played in shaping how these interactions transpired, lending equal agency to both parties.

A third thread of reasoning was that of people having clarity on *why* they have positive or negative biases towards the macaques. They were able to reflect on past lived experiences and identify the event/s that may have resulted in their particular perceptions of and feelings towards the monkeys. Gyandev, a farmer who moved to the island fairly recently, found that he could trace his seemingly innate fear of the monkeys to two separate incidents when he was surrounded by large male macaques in the forest patch near his house. A woman, Vyjayanti, found that she despised the macaques after she was witness to them pulling off a draped saree that her neighbour was wearing while trying to chase them off her farm. On the other hand, Abdul found that he loved watching the macaques play after he was forced to keep watch over his father's farm one afternoon as a teenager. These memorable instances appeared to have predetermined their long-term outlook towards the macaques they later interacted with. Mapping such events in an individual's life could help us better understand their affective geographies and, by extension, the nature of behaviours they are likely to exhibit in the future.

Finally, we found that *logos* was, at times, invoked by people as a last resort to make sense of the as yet unexplained. While hypothesising why the macaques seemed to enjoy foraging in his field, for example, Kumaraswami said,

> In my opinion, humankind's knowledge and intelligence has reached extremely high levels. We've accomplished feats up to the moon. People are now beginning to explore newer worlds and avenues. If people can achieve so much, why wouldn't the monkeys make progress of their own too? We are people with desires—if we want to taste and eat fruits

> that come in from America or China, why won't the monkeys want to try new things as well?

Kailash Ram, another farmer, arguing along similar lines, felt that monkeys may inevitably overtake human beings in their intelligence, development and progress, of course at the will of the gods. There were, however, times when even *logos* fell short. As Ranjana, an *anganwadi* (a centre providing care for mothers and young children in a rural area) worker, said, 'People think that we should rid this place of the monkeys, since we don't know what they're thinking. But the monkeys must feel similarly about us, we'll never know.'

A Meeting of Minds

In conclusion, it appears to us that the rationales that drove the diverse settler communities of the Great Nicobar Island to perceive the long-tailed macaque in certain ways were expansive and usually internally coherent. They ranged from reasons of obligation, legality and religion to sentiment and rationality, and occasionally, combinations of these. Equally important were tacit codes and rules formed through practice. Lurking under the surface were emotive states, individual temperaments and lived experiences. But finally, rather importantly, what connected much of the interspecies interactions that we observed, was affect.

In other words, one way to understand this dynamic ecology was through the notion of ethico-aesthetics (Guattari). Here, ethos would refer to the habits cultivated by people, and perhaps by macaques, as they responded and corresponded to one another. The aesthetics, in turn, would pertain to the different capacities—innate or acquired—to affect and be affected. People then learnt to respond to macaques, just as the macaques learnt to respond to them. The religious attitudes and dispositions of the people contributed to these lifeworlds as well but were more often cultivated in practice rather than being drawn unthinkingly from the scriptures. They worked as durations, emerging in the nick of time to regulate which encounters were acceptable and which were not. Such understandings of attitudes as being tacit, dynamic and arising in practice have important implications for rethinking human–wildlife encounters beyond the typical survey instruments drawn from the quantitative social sciences and applied, often unthinkingly, to conservation.

Our arguments about affect and practice, and which forces acquire significance in any given situation or circumstance could potentially be expanded further. An important force influencing and altering the wider ecological assemblage of the island, in this case, was the tsunami. Such earthly forces as the tsunami could have had enormous affective powers (Deleuze and Guattari), materialising in different ways, within both the human and simian communities and determining the course of their subsequent interactions in profound ways. In a broader, perhaps more speculative sense, we might then turn to affect not only in ethological terms (Deleuze), but as a wider ecology cutting across bodies, populations and assemblages. Can we, therefore, now conceptualise political geology as a geographical force that originally began the process of shaping, in this instance, human–macaque lifeworlds, only to culminate later in their political ecologies?

Affect thus remains, in our opinion, the most challenging of factors to observe and document in both humans and nonhumans alike: especially in their co-constructed lifeworlds. As it is both precognitive and non-cognitive, it cannot be sought conversationally or through ethnographies alone. It would appear almost impossible for one to articulate the immediate affective states of any being; any attempt at identifying them through interviews, in the case of humans, may thus be at best, the first step. It is, nevertheless, critically important to make all efforts to capture people's and macaques' affectual geographies as a means to cut through the cloud of influencing factors and to understand them for who they are, as they *are*. And as we have speculated above, affect can also emanate from earthly forces with bearings on both ethos and aesthetics. Emotive states too can be confusing, to both researchers and to those who possess them. Yet, human–nonhuman interactions are fundamentally complex and emotionally heavy, making it imperative to understand the emotional motivations that drive and influence them as well.

Finally, it is important to remember that while it may be possible to disentangle the *ethos*, *pathos* or *logos* from people's independent accounts of their experiences of living with macaques—as we have attempted to do rather preliminarily—they do not seem to separate from one another as easily in the minds of the people in the everyday. They coexist and interact—remaining entangled in often undecipherable ways—just as humans and nonhumans alike struggle to make sense of the complex world of interspecies interactions and communication within which they remain, often helplessly and inextricably, embedded.

Ethics Statement

This study was approved by the Internal Human Ethics Review Board of the Centre for Wildlife Studies—Wildlife Conservation Society, in association with that of the National Centre for Biological Sciences, Bengaluru. The names of all the respondents, mentioned in this chapter, have been altered to protect their anonymity.

Acknowledgements

We thank the National Centre for Biological Sciences and the Centre for Wildlife Studies—Wildlife Conservation Society, both in Bengaluru, for all the logistic and institutional support that made this study possible. We also thank the Sir Dorabji Tata Trust and Department of Atomic Energy for funding this work. Anindya Sinha owes a great debt of gratitude to Ambika Aiyadurai, Arka Chattopadhyay and Nishaant Choksi for their immense patience with occasionally errant authors. Most importantly, we extend our sincere gratitude to the people and the macaques of the Great Nicobar Island who let us into their lifeworlds.

Works Cited

Anderson, Ben. "Becoming and Being Hopeful: Towards a Theory of Affect". *Environment and Planning D: Society and Space*, vol. 24, no. 5, 2006, pp. 733–52.

Arroyo Rodríguez, Victor, and Pedro A.D. Dias. "Effects of Habitat Fragmentation and Disturbance on Howler Monkeys: A Review". *American Journal of Primatology*, vol. 72, no. 1, 2010, pp. 1–16.

Aureli, Filippo, and Darlene Smucny. "The Role of Emotion in Conflict and Conflict Resolution". *Natural Conflict Resolution*, edited by Filippo Aureli and Frans B. M. de Waal. U of California P, 2000, pp. 199–224.

Baker, Lynne R., et al. "Role of Local Culture, Religion, and Human Attitudes in the Conservation of Sacred Populations of a Threatened 'Pest' Species". *Biodiversity and Conservation*, vol. 23, no. 8, 2014, pp. 1895–1909.

Barua, Maan, and Anindya Sinha. "Animating The Urban: An Ethological and Geographical Conversation". *Social and Cultural Geography*, vol. 20, no. 8, 2019, pp. 1160–180.

Chandramouli, C. *Census of India 2011: Provisional Population Totals.* Registrar General and Census Commissioner, India, 2011.

Chauhan, Anita, and Raghubir S. Pirta. "Agonistic Interactions between Humans and Two Species of Monkeys (Rhesus Monkey *Macaca mulatta* and Hanuman Langur *Semnopithecus entellus*) in Shimla, Himachal Pradesh". *Journal of Psychology* vol. 1, no. 1, 2010, pp. 9–14.

———. "Socio-Ecology of Two Species of Non-Human Primates, Rhesus Monkey (*Macaca mulatta*) and Hanuman Langur (*Semnopithecus entellus*), in Shimla, Himachal Pradesh". *Journal of Human Ecology,* vol. 30, no. 3, 2010, pp. 171–77.

de Waal, Frans B. M. "What is an Animal Emotion". *Annals of the New York Academy of Sciences*, vol. 1224, 2011, pp. 191–206.

Deleuze, Gilles. *Spinoza: Practical Philosophy.* City Lights Books, 1988.

Deleuze, Gilles, and Félix Guattari. *A Thousand Plateaus: Capitalism and Schizophrenia.* Translated by Brian Massumi. U of Minnesota P, 1987.

Despret, Vinciane. "The Body We Care For: Figures of Anthropo-zoo-genesis". *Body and Society*, vol. 10, nos. 2–3, 2004, pp. 111–34.

Elanchezhian, R., et al. "Ethnobotany of Shompens—A Primitive Tribe of Great Nicobar Island". *Indian Journal of Traditional Knowledge*, vol. 6, no. 2, 2007, pp. 342–45.

Fuentes, Augustin. "Human-Nonhuman Primate Interconnections and their Relevance to Anthropology". *Ecological and Environmental Anthropology*, vol. 2, no. 2, 2006, pp. 1–11.

———. "Ethnoprimatology and the Anthropology of the Human-Primate Interface". *Annual Review of Anthropology*, vol. 41, 2012, pp. 101–17.

Guattari, Félix. *Chaosmosis: An Ethico-Aesthetic Paradigm.* Indiana UP, 1995.

Ghuman, Sartaj S., et al. "The Role of Human Creativity in Understanding Animal Cognition". *Cognition, Experience and Creativity*, edited by Jason A. Manjaly and B. Indurkhya, Orient BlackSwan, 2015, pp. 261–71.

Gumert, Michael D., et al. "Future Directions for Research and Conservation of Long-Tailed Macaque Populations". *Monkeys on the Edge: Ecology and Management of Long-Tailed Macaques and their Interface with Humans*, edited by Michael D. Gumert, Agustín Fuentes and Lisa Jones-Engel, Cambridge UP, 2011, pp. 328–53.

———. "The Common Monkey of Southeast Asia: Longtailed Macaque Populations, Ethnophoresy, and their Occurrence in Human Environments." *Monkeys on the Edge: Ecology and Management of Long-tailed Macaques and their Interface with Humans,* edited by Michael D. Gumert, Agustín Fuentes and Lisa Jones-Engel, Cambridge UP, 2011, pp. 3–44.

Gusset, Markus, et al. "Human-Wildlife Conflict in Northern Botswana: Livestock Predation by Endangered African Wild Dog *Lycaon pictus* and Other Carnivores". *Oryx,* vol. 43, no. 1, 2009, pp. 67–72.

Hill, Catherine M. "Primate Conservation and Local Communities: Ethical Issues and Debates". *American Anthropologist,* vol. 104, no. 4, 2002, pp. 1184–194.

———. "Primate Crop Feeding Behavior, Crop Protection, and Conservation". *International Journal of Primatology,* vol. 38, no. 2, 2017, pp. 385–400.

Hill, Catherine M. and Amanda D. Webber. "Perceptions of Nonhuman Primates in Human–Wildlife Conflict Scenarios". *American Journal of Primatology*, vol. 72, no. 10, 2010, pp. 919–24.

Kremer, Louise, et al. "The Nuts and Bolts of Animal Emotion". *Neuroscience and Biobehavioral Reviews*, vol. 113, 2020, pp. 273–86.

Lorimer, Jamie, et al. "'Animals' Atmospheres'". *Progress in Human Geography*, vol. 43, no. 1, 2019, pp. 26–45.

Madden, Francine. "Creating Coexistence between Humans and Wildlife: Global Perspectives on Local Efforts to Address Human-Wildlife Conflict". *Human Dimensions of Wildlife,* vol. 9, no. 4, 2004, pp. 247–57.

Malik, Javed N., et al. "Landscape Changes in the Andaman and Nicobar Islands (India) after the December 2004 Great Sumatra Earthquake and Indian Ocean Tsunami". *Earthquake Spectra,* vol. 22, no. 3, 2006, pp. 43–66.

Manral, Upma, et al. "Human Wildlife Conflict in India: A Review of Economic Implication of Loss and Preventive Measures". *Indian Forester,* vol. 142, no. 10, 2016, pp. 928–40.

Marchal, Valerie, and Catherine Hill. "Primate Crop-Raiding: A Study of Local Perceptions in Four Villages in North Sumatra, Indonesia". *Primate Conservation,* vol. 24, no. 1, 2009, pp. 107–116.

Massumi, Brian. *Parables for the Virtual: Movement, Affect, Sensation*. Duke UP, 2002.

Mishra, Partha Sarathi, et al. "Chaos in Coexistence: Perceptions of Farmers towards Long-Tailed Macaques (*Macaca fascicularis umbrosus*) Related to Crop Loss on Great Nicobar Island". *Primate Conservation,* vol. 34, 2020, pp. 175–83.

O'Grady, Nat. "Geographies of Affect". *Oxford Bibliographies*, 2018, DOI: 10.1093/OBO/9780199874002-0186.

Paul, Elizabeth S., and Michael T. Mendl. "Animal Emotion: Descriptive and Prescriptive Definitions and Their Implications for a Comparative Perspective". *Applied Animal Behaviour Science*, vol. 205, 2018, pp. 202–09.

Pile, Steve. "Emotions and Affect in Recent Human Geography". *Transactions of the Institute of British Geographers,* vol. 35, no. 1, 2010, pp. 5–20.

Porwal, Mahesh C., et al. "Impact of Tsunami on the Forest and Biodiversity Richness in Nicobar Islands (Andaman and Nicobar Islands), India". *Biodiversity and Conservation*, vol. 21, no. 5, 2012, pp. 1267–87.

Premack, David, and Guy Woodruff. "Does the Chimpanzee have a Theory of Mind?". *Behavioral and Brain Sciences,* vol. 1, no. 4, 1978, pp. 515–26.

Radhakrishna, Sindhu. "Primate Tales: Using Literature to Understand Changes in Human-Primate Relations". *International Journal of Primatology,* vol. 39, no. 5, 2018, pp. 878–94.

Richard, Alison F., et al. "Weed Macaques: The Evolutionary Implications of Macaque Feeding Ecology". *International Journal of Primatology,* vol. 10, no. 6, 1989, pp. 569–94.

Saini, Ajay. "The Southern Nicobar Islands as Imaginative Geographies". *Social Change,* vol. 46, no. 4, 2016, pp, 495–511.

Saraswat, Raghav, et al. "A God becomes a Pest? Human-Rhesus Macaque Interactions in Himachal Pradesh, Northern India". *European Journal of Wildlife Research*, vol. 61, no. 3, 2015, pp. 435–43.

Schilaci, Michael A., et al. "The Not-So-Sacred Monkeys of Bali: A Radiographic Study of Human-Primate Commensalism". *Indonesian Primates*, Springer, 2010, pp. 249–56.

Singh, Mewa. "Management of Forest-Dwelling and Urban Species: Case Studies of the Lion-Tailed Macaque (*Macaca silenus*) and the Bonnet Macaque (*M. radiata*)". *International Journal of Primatology*, vol. 40, 2019, pp. 613–29.

Singh, Vikram, and Murari L. Thakur. "Rhesus Macaque and Associated Problems in Himachal Pradesh—India". *Taprobanica*, vol. 4, no. 2, 2012, pp. 112–16.

Sinha, Anindya, and Kakoli Mukhopadhyay. "The Monkey in the Town's Commons, Revisited: An Anthropogenic History of the Indian Bonnet Macaque". *The Macaque Connection: Cooperation and Conflict between Humans and Macaques*, edited by Sindhu Radhakrishna, Michael A. Huffman and Anindya Sinha, Springer Science + Business Media, 2013, pp. 187–208.

———, et al. "Affective Ethnographies of Animal Lives". *A Research Agenda for Animal Geographies*, edited by Alice Hovorka, Sandra McCubbin, and Lauren Van Patter, Edward Elgar Publishing, 2021, pp. 129–46.

Solomon, Daniel A. "Interpellation and Affect: Activating Political Potentials across Primate Species at Jakhoo Mandir, Shimla". *Humanimalia*, vol. 8, no. 1, 2016, pp. 1–34.

Southwick, Charles H. and M. Rafiq Siddiqi. "Status, Conservation and Management of Primates in India". *ENVIS Bulletin: Wildlife and Protected Areas*, vol. 1, 2001, pp. 81–91.

Srinivasaiah, Nishant M., et al. "The Rurban Elephant: Behavioural Ecology of Asian Elephants in Response to Large-Scale Land Use Change in a Human-Dominated Landscape in Peri-Urban Southern India". *New Forms of Urban Agriculture: An Urban Ecology Perspective*, edited by Harpreet Kaur and Jessica A. Diehl, Springer Nature, 2021, 289–310.

Umapathy, Govindhaswamy, et al. "Status and Distribution of *Macaca fascicularis umbrosa* in the Nicobar Islands, India". *International Journal of Primatology*, vol. 24, no. 2, 2003, pp. 281–93.

Velankar, Avadhoot D., et al. "Population Recovery of Nicobar Long-Tailed Macaque *Macaca fascicularis umbrosus* following a Tsunami in the Nicobar Islands, India". *PLOS One,* vol. 11, no. 2, 2016, e0148205.

TWO

HOW TO BECOME WITH A BIRD

LESSONS FROM SALIM ALI'S *COMMON BIRDS*

Krishnanunni Hari

Common Birds by Salim Ali and Laeeq Futehally (1967) is one among several texts constituting the complex discourse that determines our apprehension of animals in India. Depending on the species, its cultural, political and commercial significance and its location, these texts manage the animal body appropriately for human consumption, fascination or 'indigestion' (Haraway 285). This objective and scientific guide for birdwatchers—as opaque as can be—shows interesting symptoms upon close reading signalling that something more is going on than what first meets the eye. While the implied author–birdwatcher figure embedded in the text legitimises only some ways of being with birds—as a *shikari*/hunter, as an ornithologist collecting specimens, as a scientist, or as a farmer exterminating pests and encouraging other birds which are beneficial to him—his risky attentiveness to bird life endows *Common Birds* with an ethical and political purchase in imagining other relations. This both complicates and exceeds the text's professed instrumental intentions.

Setting the Stage: The Preferred Reading of *Common Birds*

The introductory paratexts of *Common Birds* and Ali's scientific and economic standpoint regarding ornithology, *shikar* (hunting) and conservation as discussed in his autobiography, *The Fall of a Sparrow* (1985), prescribe the manual's preferred reading as simply a field-companion to birdwatching. For instance, Ali de-emphasises both a 'prick of conscience' at killing birds and his own scepticism regarding man's superiority over his fellow animals, '... a dubious superiority over lesser creatures', in the interest of furthering scientific knowledge. Similarly, his scruples at shooting birds for collection ('several thousands, alas') and his conviction that the only difference between man and other animals is the former having a more highly evolved brain, goes together with his non-sentimental 'aesthetic and ... pragmatic' love for birds. He deems *shikar* as essential to regulate unchecked prey populations—they may be culled through sports-shooting and utilised as a protein-rich food resource. He reduces wildlife conservation to '... practical ... scientific, cultural, aesthetic, recreational and economic reasons [with] sentimentality [having] little to do with it' *(The Fall of a Sparrow* 195, 233). Ali states that the primary reason behind writing *Common Birds* and for pursuing birdwatching is economic. It is 'significant and urgent' in a forested, agricultural and thickly populated country like India to distinguish man's friends from his foes; the 'social organization, population dynamics' and how they bring up their families are all relevant insofar as they contribute directly or indirectly to human economic progress (*Common Birds* 8).

Common Birds also highlights the trends in ornithology that emerged in the 1960s. Ali calls for a transition from collecting and rendering live birds into 'dry museum skins' to studying the 'life-history' of different bird species (8). Defined by Merriam-Webster Dictionary as 'The series of changes undergone by an organism during its lifetime', the idea of life-histories for birds begins to unsettle the standpoints that have been erected in the manual. Forming the entirety of its main body, life-histories demonstrate first a collapse of the human–animal binary.

FOREGROUNDING THE ANIMAL BODY

Jacques Derrida in 'The Animal That Therefore I Am (More to Follow)' locates the original act of violence against animals at the moment when man, sinning against all 'empirical authority', 'corrals' the entirety of nonhuman animate life into a general category that he calls 'The Animal' (124). It is co-produced with an abyss separating human animals from nonhuman ones which, at once, institutes a hierarchy on the arbitrary basis of *logos*, or language, as proper only to man. In essence, one can think of the lack of *logos* in 'The Animal' as nothing but 'a privation' (126). Never a subject and therefore never responding but only reacting, 'The Animal'—a vague grouping of all the diverse forms of life which do not meet the historically contingent checklist of being Human—is reduced to being identified by and with this lack (125).

Birdwatching's reliance on classification helps to avoid, to some extent, reproducing the logics of homogenising differences between nonhuman animals. Classification distributes a singular difference across a field of differences like in the shape of beak and claws, nesting and social habits, among others. Classification, as it is performed in the writing of life-histories, brings into focus the lively variety within the seemingly uniform category of the animal and the human–animal binary that it sustains. Life-histories depend on the observation of everyday pursuits and occurrences animating individual birds in the here-and-now. This exercise of building knowledge through species stories by its very 'ongoingness' reminds us how birds do not follow a predictable behavioural script, but uniquely respond to the meaningful changes in their environment. Furthermore, writing histories of everyday bird life displaces the idea of history as a prerogative of man, the 'autobiographical animal' (Derrida 126). The birds enact their agency by affecting the birdwatcher with their animatedness, which he translates into the text; one can say that bird and birdwatcher write the life-histories as one. Before making that argument, it is necessary to look into what this animatedness is made of.

Confronted with the stupidity or '*bêtise*' of the human–animal binary, Derrida introduces two linkages that situate human and nonhuman life in continuity with each other (124). The first is through an analysis of Jeremy Bentham's question, rephrased as, 'the question is not to know whether the animal can think, reason or talk ... The *first* and *decisive* question will rather be to know whether animals *can suffer*' (121). Derrida sees the very 'protocol' of this question as pulling the history of thinking out of its stagnation in an endless accounting of what is in the animal's power or capacity to do and have. The animal, who is essentalised by and as lack, fails the test even before it has started. Bentham's rhetorical question thus shifts the terms of the 'first question' from 'being able' to 'can they *not be able*'; it stands testimony to the 'undeniable[-ity]'of the 'nonpower' of suffering in all finite mortal beings (121).

For his second move Derrida opens up 'trace' as that which emerges in movement and which is not limited to language. 'Auto-motricity', he writes, '[is], a spontaneity that is given to movement, to organizing itself and affecting itself, marking, tracing and affecting itself with traces of its self ...' Marking animality as distinct from inorganic matter, auto-motricity precedes *logos*. Derrida's response to the identification of animals with a lack is hence, through a demonstration of how they always already autobiographise themselves—'... *it itself is* ...'— by means of a trace that is inscribed by any mobile, vulnerably embodied being (127). This self-sufficient subjecthood obviates the need for a supplementary logocentric articulation. A similar conclusion is arrived at by Elizabeth Costello, the novelist–protagonist of J.M. Coetzee's *The Lives of Animals* (2016): the self-containment of our animal being consists in the 'joy' of living as a 'body-soul'. She develops this thought in a reading of 'The Jaguar' by Ted Hughes: 'We know the jaguar not from the way he seems but from the way he moves. The body is as the body moves' (33).

IMMERSIVE ANIMALITY AND ATTUNEMENT

An animal whose mind is not split from its body, whose 'consciousness is kinetic ...' (Coetzee 33), autobiographises its being through spontaneous movement. This animatedness of the bird/animal as copied out in language by the birdwatcher features moments of Ali's immersion in the bird's embodied becoming. This is felt in the exhilarating pace and wording of some of the descriptions. More than just being vivid, they suggest new ways of relating with birds. The following instances of the little cormorant and the Bengal vulture illustrate my point:

> Sometimes they hunt by concerted action, a flock hemming in a school of fish and chasing and diving repeatedly after the quarry, jostling and leap-frogging over one another in feverish excitement (*Common Birds* 24).
>
> These gruesome obsequies are attended by incessant jostling and bickering among the feasters and much raucous screeching and hissing ... [they] prance around with outspread wings, tugging and pulling at a gobbet of flesh from either end (*Common Birds* 41).

Additionally in *The Book of Indian Birds* (1996), the common iora is seen, '... springing several feet up in the air, fluffing out and exhibiting the glistening white feathers on his rump and parachuting down to his perch in spirals looking like a ball of fluff' (24). The verbs—chasing, diving, jostling, tugging—seem to assume a bird-life of their own.

This generation of an affect where the birds and birdwatcher merge into each other is better illustrated through the framework of attunement. Kari Weil in her critique of Heidegger's concept of 'captivation' examines how the capacity to be '... affected or tuned ... and gripped ...' by their environment constitutes the experience of animality (Weil 33). Such an attunement is 'fallen-into' outside the bounds of conscious knowledge. Heidegger uses the human ability to wake up from being captivated in the environment and reflect on the world 'as such' to separate the human from the animal. However, humans in being unknowingly and '... constantly drawn outside oneself to the environment and to the Other in the environment ...' remain rooted in animality. Ali's 'intense investiture' in birds' becoming-in-motion is liberated from an obsessive calculation of the economic benefits of each bird precisely because he is not aware of being attuned (Weil 43). The birdwatcher indulges in the immersive sensory world of animality within and between the bird and himself. Overlaps such as these, where the centrality of the human subject and his symbolic knowledge of the world are shaken during an attunement to avian kinesthetics, make up the ground for imagining a non-hierarchical ordering of human–bird relations.

ATTUNEMENT FOR CO-CONSTITUTION OF TEXT AND BODY

Vinciane Despret in her paper on redefining domestication presents a different sense of attunement. Attunement here becomes a practice that induces 'new articulations' between consciousness, affects and bodies. Looking into two case studies—Hans, the clever horse who seemed to be able to count and the psychologist Rosenthal's experiments with his students—Despret traces a network of trust and faith on and between each interested party that enables them to propose 'beliefs' and fulfil each other's expectations (114). For instance, Rosenthal, by (mis)leading his students to prove that neutral rats (mis)labelled as bright are actually bright, tries to 'authorise' a certain way of performing experiments. He makes the point that the experimenter must recognise and avoid subjective influences from his objects of study as errors distorting objective reality (120). Instead, the students actually believe, and thereby authorise, their rats to be bright. The animals 'make [themselves] available' for this proposed brightness; they eventually make good on the students' belief, producing themselves as bright rats and the students as 'competent' in recognising bright rats (114).[1] One observes how each party becomes with their companions in multiple ways that both meet and challenge the results expected by authority figures.

Despret's insights into attunement can be productively engaged with in *Common Birds*. The manual is delimited by the disciplinary requirements of birdwatching which tries to justify its relevance by relying on an ability to contribute to economic progress—birdwatching is pitched as building knowledge about birds which aid human endeavours like farming and those which are inimical to the same. Consequently, Ali himself authorises a particular purpose and method for writing life-histories that are grounded in economic utility. The fact that there is certainty about these positions—which also shape the overall structure of the book—is enough to infer that the birds and birdwatcher have complied with authorial expectations. The birds live lives that are malleable enough to be recast as a means to anthropocentric ends while the birdwatcher observes his subjects with the required objectivity—they are made to produce each other only in required ways.

These rather neat, complementary authorisations are also resisted and exceeded to give richer co-composed accounts of bird agency. The birds, as one may gather from the excerpts above, refocus Ali's attention to their lives, making his body 'be moved' by their liveliness. The same bird bodies that he has produced through attunement as affecting, in turn produce him as an affected body allowing him to make vibrant accounts of birds' kinetic spontaneity. They tune him in such a way that he becomes competent to recognise and respond to their liveliness—in short, he becomes response-able (113). Such careful crafting from both sides is possible because both parties have made themselves available, '... at the fringe of consciousness', to an attunement to each other as well as to their environment at large (qtd in. Despret 114). The descriptions, '... articulated [as they are] by affects', become a record of this transformative engagement or entanglement, a record that is co-composed by bird and birdwatcher (Despret 125). Through his now competent and responding body the birds write themselves into the text in a dance of 'companion species' that co-produces both textual and corporeal bodies (Haraway 32). We observe how the arts of attunement create a 'setting' within the larger framework of birdwatching that mostly disarticulates birds. This resistant former setting counters the predominant order and the hierarchised identities that it institutes, opening up the possibility for other articulations through richer accounts of subjects-in-the-making (Despret 131).

PLAY AS EMBODIED POSSIBILITY

Becoming otherwise is a potentiality contained amply in play. Donna Haraway suggests something similar in her book, *When Species Meet*, when reflecting on the agility games she plays with her dog, Cayenne. Communication is embodied in play; action is moved between contexts and new and risky relations forged with one's companions. Following Mark Bekoff, play is recognised as the 'purposeless' use of 'motor activity' in altered form and sequence within contexts where they do not functionally belong. They serve no other function but to play for the 'sheer joy' of it; 'it discloses living' (240). Yet, in the life-history of the Bengal vulture and little cormorant—where the birds indulge in the very functional acts of pulling and tugging shredded carrion or trapped quarry—play emerges as 'the open'—'never certain, never guaranteed'—where new embodiments are proposed and realised (368). The present progressives—'jostling', 'prancing', 'leap-frogging', 'diving'—show a significant absence of markers separating one action from another. They merge into each other and combine in new ways to delineate new forms for the body. 'Play proposes' in the moment of passionate necessity. It opens up prospects for invention by rearranging actions and their temporal sequences (240).[2] This frenzied action co-produces a confluence of bodies in co-shaping motion with each other—a coherence of quickly emanating, quickly fleshed-out configurations.

Unlike Haraway's agility partner, the individual vultures and cormorants are strangers to Ali. Neither is their becoming with their conspecifics specifically addressed to him. Instead, they cohere into a 'chimerical', multiplicitous 'figure' that collects others in their surroundings into the intense flows of a shared physical thrill that affects all (*Common Birds* 4).[3] These new entanglements materialise 'outside time' or in 'an eternal present' (Haraway 240). The present progressives stand suspended between the completedness of the simple past and the measurable regularity of the simple present. 'Chasing', 'diving', 'springing'—all occupy times of their own. The birdwatcher is literally attuned to the 'harmony' of the birds' becoming; the time in which this harmony expands undoes his own experience of 'Time' (Buchanan 26).[4] Furthermore, it must be remembered that the joyous commingling of bodies and affects is never a silent phenomenon; it comes with its own wall of sound. In witnessing with the whole sensorium, an immersive attunement displaces the ascendancy of sight from the practice of birdwatching, unsettling also 'Sight's' 'ineluctable ties' with the 'repression' of animality, 'valorisation' of the human and distancing the subject from his physical environment (Wolfe 3). Throughout *Common Birds* and *The Book of Indian Birds* Ali also onomatopoeically transcribes the calls of different birds, like 'the subdued *kroo-kroo*' of the Deccan scimitar babbler for instance (18). At once mimicry yet more, these renditions—the single '*chip*' of the streaked fantail warbler compared to 'the snip of a barber's scissors heard in the distance' being one example—emerge as a creative re-performance that his new identity of birdwatcher-with-bird enables. Ali simultaneously posits a new way of being birdwatcher and '... a new way of being human-with-nonhumans' (*Indian Birds* 86; Despret 131).

It is not only the spectacular moments of bird liveliness to which arts of attunement are crafted in *Common Birds*. A deep attentiveness to the everyday doings of birds—nesting, feeding, socialising and courting mates—creates the stories that form a major portion of the life-histories. As we have already seen, tuning freely into the affects that animate bird bodies has consequences for flattening hierarchies, reclaiming the experience of animality and imagining relations that allow all subjects to affect and alter each other responsibly. Similarly, an attention to minute details of the birds' 'phenomenal worlds', or *Umwelt,* generates knowledges of the factors that play a role in their flourishing (qtd in. Weil 11). Birdwatching practices thereby serve to 'body forth' the birds, moving them from being abstract, unknowable or insignificant forms at the margins of our lives to unique, individual members entangled with us in non-innocent relations (Coetzee 53). To learn more is to care more about what matters to the birds themselves; it means having stakes in readjusting our spaces and lives so as to maximise their opportunities for co-flourishing.

The authorised *raison d'etre* of the manual, however, predetermines what will be looked for in bird lives and why. For each species—and species classification itself, as Alexis Shotwell reminds us, indicates Man's dominion over Nature through his 'God-given right to name' and rationalise—we thus have a balance sheet of their benefits against their harms for human society (98). Even though economic value and instrumental rationality inform the organising principle of the manual, the arts of attunement that it relies on make sure that its original intentions are always surpassed by a 'paradoxical' interest in the uniqueness of the birds themselves (Shotwell 98).

One observes such a takeover in Ali's playful depictions of the everyday social life of birds. For example, in 'sisterhoods' of the jungle babbler, 'occasionally differences of opinion arise between members, and loud and discordant wrangling ensues'; the 'whole sisterhood' of the common babbler combines '... to hurl invectives at [the intruder] in disorderly chorus'; Jerdon's chloropsis '... acts the blustering bully, attacking and driving off every other bird ... [they] will even resort to dog-in-the-manger tactics ...' *(Common* Birds 14–16; 28). The birdwatcher's attunement to the birds makes them 'available' in such a way that they propose a parallel—or something that is readily recognised as a parallel to his own *Umwelt* (Despret 114). Propositions like 'holding opinions', 'hurling invectives' and 'bullying' are enacted by the bird's body which has been produced to do more than its instrumental function. Reciprocally, the birdwatcher is made available to receive and reinforce a reality where birds are seen by humans as hurling invectives.[5] These playful, immersive adventures open up society and culture as more-than-human phenomena. Attunement holds the birds and birdwatcher not as essentialised, autonomous identities, but in relations of becoming more knowledgeable of and with each other. Further as Despret tells us, learning to 'address' birds as full, agential subjects—which birdwatching practice, despite its teleological intentions, teaches us to do—forms the primary 'condition' that has to be met for even instrumental life-histories or ornithological theory to be formulated (131).

THE POLITICS OF ATTUNEMENT

The motive behind writing life-histories, Ali declares, is economic. This economic orientation, as it plays out in the text, maps the different non-innocent relations of use—as pollinators, scavengers of dead cattle, predators of insects harmful to agriculture—between humans and animals in India. By putting each other to use, birds and humans create meaningful overlaps between the instrumental use of others and mutual flourishing with them. Entanglements of relations of use where humans have unequal power to decide the fates of other animals, where suffering is differentially distributed and mostly exported to animal bodies, are the 'contradictory truths' of worldly existence (Haraway 102). At the same time, animals as agential subjects, constantly looking back at us, are entitled to real opportunities for their self-actualisation within these power saturated relations. To meet this obligation, it is necessary to rethink our relations so as to prevent the de-subjectification of animals and to nourish practices of 'nonteleological care' (Shotwell 104).

In Ali's texts a predominantly instrumental outlook is also countered by the birdwatcher's attunement that unfurls avian agency. However, this attunement does not apprehend all birds with equal openness. In startling contrast with the babblers who are affectionately described, the common house-crow is contemptuously characterised in moralistically heavy expressions—'. . . a life of sin and wrongdoing'; 'thieving propensities', among others (*Common Birds* 3). The birdwatcher's reception and interpretation of the proposition to see the crow as more than the function it serves is invariably political; it is inflected by and embedded in an 'institution of speciesism' (Wolfe 7).[6] Consequently, the crow's ingenuity, prudence and boldness are immediately recast as sinful and wrong. The same attunement that indulges the babbler's resistance to the authorial framework makes the crow 'docile' under the weight of an uncompromising instrumentality. It is begrudgingly tolerated or 'redeemed' through a concluding tally that sums up its status as 'neutral'. The case of the house crow demonstrates how care does not inhere in relations of use. A logic that reduces birds to their use for humans also makes the continuation of their lives contingent on the non-availability of alternate tools to profitably perform the same function—it is solely the crow's role as an effective scavenger that prevents its classification as vermin. Responsibility within instrumental entanglements must emerge as a non-speciesist regard for the birds as worthy in themselves.

The ethical and political significance of the arts of attunement consists in re-calibrating a scientific method of observation and knowledge production, well-suited as it is, for affecting and being affected by individual members of avian species. This allows one to imagine and work towards worlds other than those that an objectifying gaze of economic instrumentality exhaustively structures as the inevitable and only possible way of being and becoming. Ali's immersive birdwatching practices redistribute fixed identities of the human and the animal as particularised co-produced encounters unfolding in entanglement with each other. Such an event can be appreciated only when we take into account the humans and birds gathered by the text into the practices of attunement that *Common Birds* embodies. Serving as a field guide for current and prospective (often lay) birdwatchers who answer Ali's call to sketch the ongoing life-history of birds in India, the text grows and multiplies with the accrual of each new account, edition and adaptation. The multiplicitous body of texts that *Common Birds* comes to represent, in effect, is in becoming with multispecies communities of attunement and care who co-compose a testimony of bearing witness to the world in times of biosocial destruction.

Conclusion: Attunement and Liberation

Practising attunement makes the birdwatcher competent to recognise and co-articulate previously silenced avian agencies. Tuning into another's world and attending to what matters to them institutes a responsibility to work toward structures that ensure their self-actualisation. However, there is no clear conception as to what fulfils this responsibility. Borrowing from Haraway, any responsibility that is worth its name must be open—not just to simple tweaking, rearrangement or subversions of instrumental relations—but also to the equal possibility of annihilating relationships that are predisposed to unilateral power and oppression. Not making, say, a pig that is bred, raised and marked for slaughter 'killable' involves taking the pig's affective response seriously enough to allow it to undo the discourse that ontologises the pig as always already meant for meat (Haraway 80). Responsibility commits to equally enhancing the other possibilities for porcine becoming. In contrast, the general ambiguity around responsibility in power-fraught entanglements leads to a situation where its demands are met even by the rather low standards produced by powerful human discourses that have an interest in trapping the animal in its objectified status.[7]

Attunement as an ethico-political strategy is thus insufficient by itself in dismantling oppressive relations and attaining better worlds. Krithika Srinivasan, in a paper on the biopolitics of dog control, evaluates both the effectiveness and inadequacy of relational frameworks like attunement in understanding human–animal relations. Relational approaches, she observes, play an important role in foregrounding animal agency in relationships between individual humans and animals, thereby challenging 'dualist' and 'human exceptionalist' ontologies. However, the often 'purely human constructs and decision-making' render insignificant the influence that animals have on the 'human discourse' that legislates on their life and death (2). A comprehensive liberatory framework would thus complement the agency-articulating effects of attunement with top-down structural and societal change through analyses and modes of action that are not 'contingent' on the specificities of individual relationships (3). It would be liberatory inasmuch as it 'opens freedom' to and in community with human and nonhuman others (Shotwell 128).

Notes

1. Despret follows Gregory Bateson's definition of 'Authority'—'a person is said to have authority when anyone who is under the influence of

that authority does everything possible to make whatever this person says be true' (qtd in. Despret 119).

2. Propositions are, Haraway explains, that which can be said to exist. The coming into existence of propositions '... opens a full range of new diverging possibilities for becoming ... and as such generally signifies a break in continuity' (243). On the play of propositions, see Haraway's discussion of Stengers and Whitehead (243).
3. As Haraway explains, 'figures' are 'chimerical visions'—material–semiotic nodes where the heterogeneous bodies and meanings that they collect co-shape one another (4).
4. Uexkull's 'Harmony begins with at least two different organisms acting in relation with one another ... harmony can also extend to a collective whole, such as a swarm ...' (Buchanan 26).
5. Contrary to the impoverishing effect of the term, anthropomorphising—a charge that can be levelled at Ali—here 'adds new identities' as his body is being produced by a new identity, Ali-with-birds (Despret 130).
6. The institution of speciesism is defined by 'the ethical acceptability of the systematic "noncriminal putting to death" of animals based solely on their species' (Wolfe 7).
7. On the poor standards of responsibility when it is adapted into the business of meat production, Ian Hacking writes in 'Deflections', '[Temple] Grandin's measure of success, in making a slaughterhouse more "humane," is that *only one in four* of the victims becomes so frightened that it needs to be driven to its end with electric cattle prods' (Cavell et al. 150)

Works Cited

Ali, Salim. *The Book of Indian Birds*. Oxford UP, 1996.

———. *The Fall of a Sparrow*. Oxford UP, 1985.

———, and Laeeq Futehally. *Common Birds*. National Book Trust, 1967.

Buchanan, Brett. *Onto-Ethologies: The Animal Environments of Uexkull, Heidegger, Merleau-Ponty, and Deleuze*. State U of New York P, 2008.

Cavell, Stanley, et al. *Philosophy and Animal Life*. Columbia UP, 2009.

Coetzee, John Maxwell. *The Lives of Animals* [Princeton Classics], U of Princeton P, 2016.

Derrida, Jacques. "The Animal That Therefore I Am (More to Follow)". *Animal Philosophy: Ethics and Identity*, edited by Peter Atterton and Matthew Calarco, Continuum, 2004, pp. 113–28.

Despret, Vinciane. "The Body We Care For: Figures of Anthropo-zoo-genesis". *Body & Society*, vol. 1, nos. 2–3, 2004, pp. 111–34.

Haraway, Donna J. *When Species Meet.* U of Minnesota P, 2008.

Shotwell, Alexis. *Against Purity: Living Ethically in Compromised Times.* U of Minnesota P, 2016.

Srinivasan, Krithika. "The Biopolitics of Animal Being and Welfare: Dog Control and Care in the UK and India". *Transactions of the Institute of British Geographers*, vol. 38, no. 1, 2013, pp. 106–19.

Weil, Kari. *Thinking Animals: Why Animal Studies Now?* Columbia UP, 2012.

Wolfe, Cary. *Animal Rites: American Culture, The Discourse Of Species, and Posthumanist Theory.* U of Chicago P, 2003.

THREE

FEELING THE MOSS

EXPLORING ECOLOGICAL SENSITIVITY THROUGH 'ENCHANTED' ENCOUNTERS

Deborah Dutta

'We have become experts in analysing what nature can do for us, but lack a language to evoke what it can do to us'—this critique by Robert Macfarlane (24) is not unfounded, if one were to observe the dominant discourse regarding nature. Knowledge about the environment is not new. In fact, in evolutionary terms, humans have always had an intimate understanding of the environment, as we have spent most of our time on the planet foraging and hunting in forests or engaging in farming. However, different from this knowledge acquired through participating in the environment, current forms of Environmental Education (EE) provide knowledge about the environment. This descriptive approach within the formal education system is largely a crises-driven modern agenda, which came into being in the 1960s when some scientists turned their attention to ecological problems caused by techno–scientific enterprises. A major focus of these nascent environmental movements, when translated into educational endeavours, was creating public awareness of various ecological problems.

Indian thinkers had paid attention to environmental education even before the environment become a topic of interest at the international level from a different route. This was partly due to India's particular geography, but sociohistorical events such as the British occupation and the freedom struggle also significantly contributed to this interest. Initial efforts to formally include environmental practices within education were seen in Gandhi's formulation of education (Sinha 96), known as 'Nai Taleem', which stemmed from his understanding of local community and family as the basic unit of sustenance and resilience. Though conceived as a resistance movement against the colonial system of education—which was aimed at creating employees for the British administration—these educational interventions focused on inherently ecologically sustainable practices. However, post-Independence education policies moved away from ideas such as 'Nai Taleem'. This shift began with the government's focus on 'modernisation' through large-scale science and engineering projects/ institutions ("Quality of Education" 10). This way of thinking is etched in policy documents, particularly throughout the 1966 policy document on education ("Agricultural Modernisation" 2370), where a permanent wedge is cast between core disciplines (such as math and science) and 'extra-curricular' subjects such as farming, weaving and other artisan forms. The international focus on ecological issues brought back the environmental discourse in India in the late 1970s, but now from a Western, information-centred perspective.

Yet, despite access to exponentially increasing environmental information compared to past decades, there have not been many impactful actions to limit the ongoing damage to the environment. Stevenson (145) points out that formal educational systems have not been designed to promote transformative actions, as they were originally conceived as social structures to transmit cultural beliefs and cumulative knowledge gained by society. Environmental education (EE), on the other hand, requires questioning dominant cultural beliefs and developing alternative practices that are ecologically sustainable. This puts the aims of EE at odds with the general aim of schooling. An information–deficit model of learning and 'patching' environmental topics to existing disciplinary content, is not the right approach to EE as it can do little to engage with the complexity of problems, or develop skills required to act in competent ways (Almeida and Cutter-Mackenzie 333; Ashley 273; "Sustainable Development" 255).

THE EDUCATIONAL ENTERPRISE OF REANIMATING OUR SENSES

The currently dominant view of the environment is object-centered and compartmentalised, which foregrounds the idea of nature as a resource in need of better management (Capra 30), rather than the view of nature as a living web of interrelationships in which our lives are embedded. The notion of embeddedness requires encounters with the world that hold possibilities of a reciprocal relationship with nature, rather than a one-way relationship of exploitation, which derives from the logico–deductive worldview of nature as a passive entity awaiting investigation.

Intellectual distancing of actions from reasoning, considered a virtue in science, creates inherent conflicts in the environmental education discourse where appropriate actions need to stem as a natural outcome of the educational process. Secondly, equating formal science-based environmental knowledge with the knowledge for EE delegitimises forms of knowing that are implicit, experiential and context-based, as elaborated by studies of diverse Indigenous communities (Abram 26; Bird-David 99; Ingold, "Perception of the environment" 205). Dominant forms of science education focus on nurturing scientific attitudes, characterised by rationality, hypothesis and experimentation, rather than attitudes towards science which can be extended to attitudes towards the environment. The latter encompass affective experiences and motivations, which are rarely seen as important concerns while approaching a topic (Alsop and Watts 1050; Raveendran and Chunawala 670). The required primacy of action to address environmental issues calls for conceptions of EE that are different from science education, where due emphasis is given to the conative and affective domains of the human mind. Taking this critique further, I argue that close and sustained encounters with a 'rooted' practice ought to form a crucial aspect of environmental education, through providing an 'education in attention' (Ingold, "Education and/as Attention" 20). This project was motivated by a pragmatic concern regarding our collective inability to respond to the growing environmental crises in an impactful manner. How can we bring about an 'ontological transformation' (Payne 171), such that our lived reality reflects a meaningful grounding within the patterns of the ecosystem? Drawing on recent work by various scholars (Bai 147; "Environmental Concern" 286), I contend that recognising and appreciating sensory relationality with one's environment forms the core of ecological sensibilities. Through a case study of a school community farm, I explore how the everyday aesthetics of 'enchantment' (Bennett 5) allows us to see the mundane in the root sense of the word, 'as belonging to the world' in all its vivacity and life. Enchantment is a moment when the familiar is suspended from usual categorisations and can appear surreal, thereby allowing for subversive views to open up. It nurtures ways of exploration and imagination that are not amenable to language-based discourses. As a highly affectual encounter, moments of enchantment open ways of meaningful engagement with one's material environment (Pyyry 1396). These need not always be pleasant experiences (such as finding shreds of plastic in the compost, or a plant under severe pest attack) but are powerful in terms of disrupting disengaged modes of interacting with one's environment. The observations of the students at the farm showed that honing one's attention towards such particular aspects of the environment generated instances of 'response-ability' (Moriggi et al. 3; Kayumova et al. 207), wherein individuals could respond to and partake in a reciprocal sense of well-being. Their attribution of emotional states to the creatures and plants could be argued as ways of empathising with living entities as responsive and deserving of care. I conclude by submitting that the integrative form of situated and embodied interactions (such as those involved in farming), can provide the generative force to learn, act, care and live in ways that encourage ecological flourishing (Dutta and Chandrasekharan 1200).

ROOTING FOR GROUNDED ENCOUNTERS

It has been argued that students need to be exposed to affective experiences through engagement with the local environment, to foster a situated and embodied understanding of diverse ecological practices ("Childhood Experiences" 146). Within the structure of formal education, the notion of experience tends to get reduced to observations and rituals, which are episodical and uncritical. Activities meant to nurture environmental sensibilities tend to take the form of tokenistic actions such as planting saplings on Earth Day or making 'Save the Tiger' posters, without the possibility of any feedback or consequence. On the other hand, textbooks are filled with bleak scenarios of environmental degradation leaving students acutely aware of 'big' problems, but the information does not empower them to bring about any transformation in their own locality. Many educators have, therefore, argued for the need for 'authentic participation' (Jensen and Schnack 164; Hart 10; Chawla and Heft 205; Barrett 508), where students feel ownership of and take responsibility for the task at hand. In this context, Rautio (400) considers informal spaces of interaction between children and environmental artefacts as a rich source for perceiving the interdependence of relationships. She sees such interactions as a move away from the narrative of environments simply facilitating children's agency in a directed fashion. She exhorts educators 'to trust that the interaction between children and the world, seemingly irrational and mostly unreflected, also has value. This value, arguably unmeasurable, could be thought of as intrinsic and grounding' (402).

As a form of experience, food gardens are becoming an increasingly common feature of schools worldwide (Dirks and Orvis 443; Cairns 310; Blair 16). They have most commonly been employed as a measure to positively impact children's health and nutrition (Duncan et al. 765; Ratcliffe et al. 36). A different set of studies have focused on the use of food gardens as a pedagogical tool to teach natural science and other subjects (Green and Duhn 62; Passy 24). However, this utilitarian approach to food gardens has been critiqued for neglecting affective dimensions of children's relationships with the plants. Wake (423) for instance, argues that most gardening programs are usually designed and maintained by adults, with children having minimal agency in imagining and developing the space. As a result, there is less research on the processes and nature of interactions underlying farming activities (Cutter-Mackenzie 129; Dyg and Wistoft 1181).

This chapter aims to develop a process-based understanding of students' close involvement in farming activities and the potential of such activities in motivating wider ecological perspectives. More broadly, the chapter attempts to redirect attention from evaluating connectedness of two separate units (that is 'child' and 'nature') to mapping the mutual emergence of children and their surroundings in relation to each other.

Background of the Study

This chapter is based on a terrace farming project involving Class VIII students in a Central Secondary Board Education (CBSE) affiliated school at Mulund, a Mumbai suburb. The site was chosen based on availability. School spaces are usually difficult to access due to multiple levels of constraints, ranging from general suspicion towards 'outsiders' who may disrupt the public image of a school to issues of time, commitment and minimal expenses that a school has to invest during the course of any intervention. In this case, the school is located near a landfill. The visceral experience of being located near a waste dump had prompted the school to participate in many activities related to waste management and the recycling of e-waste. The principal found the terrace farming proposal interesting from the viewpoint of using organic waste to make compost and agreed to provide some space and time to carry out the activity. The discussions resulted in 40 students from Class VIII participating in the project for an hour every week. The project continued for an entire academic year.

A DESCRIPTION OF THE SCHOOL SITE: THE ROOF WITH A VIEW

'Wow! Are we going to grow plants here?!', exclaimed a student echoing the sentiments of many of her peers who rushed onto the terrace, full of excitement. The excitement mostly stemmed from just the fact that they were being allowed to work in a space that was usually locked out of view. The terrace was completely barren and offered a good view of the city landfill that could be mistaken for a hill, having a decent green cover during the monsoon season. The students' reactions ranged from curiosity to apprehension. The year-long project was deliberately kept open-ended, with the broad goals of growing plants using principles of organic farming while using minimal resources. The project was not graded, thus encouraging the students' autonomy and freedom to engage with the project based on their interest. Every session would typically start with some observation of the surroundings, followed by a discussion and suggestions for activities. Students usually worked in groups of 3–4, often forming even larger groups if they felt the need (for instance, activities like trellis-making required a large group to hold and tie the structure together). Major activities in setting up the farm included making a compost pit on the school grounds, collecting dried leaves to add to the compost and mulch the plants, making sapling planters, making cardboard planters, making supports for climbers and creepers, plant care, saving seeds and harvesting. Occasionally, a liquid called *Amrut Jal*, made from a mixture of water, cattle urine, cattle dung and jaggery was added to the plants to help microbial growth of the soil. Cattle dung contains many microbes that aid in decomposition while the urine has high amounts of urea, which creates an ideal ambience for the microbes to multiply. The jaggery aids in fermentation. The method harks back to traditional practices of keeping cattle near the farm, thereby allowing a mutually beneficial relationship between the soil, farm produce, cattle and the farmer to emerge. The students also made a batch of *Amrut Mitti*,[1] made by decomposing dry biomass (comprising mostly of dry leaves using *Amrut Jal*), eventually.

Most students in the school had grown up in cities and had fairly limited ideas about growing plants. Some had a few ornamental plants at home, but the idea that edible plants and vegetables could be grown in a small area was new for most of them. Students were allowed to explore, observe and play while participating in various activities. As a result, students had varied perspectives and motivations that evolved along with the setting up of the terrace farm. Some students were initially unwilling to get their hands into 'dirt' and preferred observing others. Some others had a more scholarly interest in the process and would ask questions about the types of decomposition, how much time it could take, why didn't some seeds sprout and so on. As a co-facilitator, I allowed for multiple modes of participation. Students would participate in whatever activity they felt most comfortable doing.

EVOLVING SOMAESTHETIC INTERACTIONS

Sensory participation was central to the students' experience of the terrace farm. Students were observed engaging with plants in a rich, visceral manner through senses of touch, smell and taste, thus widening their modalities of perceiving the environment. The visceral sensations of tasting the plants, digging the soil, stroking the leaves, gingerly handling the seedlings, feeling the movement of insects on their fingers, smelling the composted soil and countless other encounters—particularly in relation to the growing of plants—'invited' students to participate in an evolving, reciprocal relationship with the farm environment. Engaging in different modalities of perception (as opposed to a dominant visual mode) facilitated what Abram (12) described as a shift from description 'about' to correspondence 'with'—students were responding to the plants rather than studying them. Iared et al., argue that 'close contact with the unpremeditated sensible world awakens in us biophilic feelings, precisely because we have a common origin with the elements of nature' (194). The natural tendency to explore and bond with the natural world is termed as biophilia. Scholars have argued that for children's natural inclinations to develop, appropriate and immersive nature-based initiatives are required (Sobel 6; Chawla, "Childhood Experiences" 150).

To illustrate, students had never seen the plant called Indian roselle (locally called *ambadi*). It was grown on the terrace and they were informed that the leaves of the plant are edible. Initially, for most students, the mere idea of eating something directly off a plant was a novel concept, given that their interaction with food was mostly in packaged, frozen or cooked form. However, apprehensions gave way to curiosity as they began to sniff, taste and finally nibble the leaves tentatively. The sour tasting leaves went on to become a garden favourite, as evident by frequent comments like, 'The leaves taste so sour ... And I liked to eat it!'.

> We used to come every day, excited for terrace farming but the main reason was that we would get to, used to actually eat the plants. There was that *ambadi* plant, it was sour and we also actually opened up many of those, you know, the containers not containers actually, but the parts in which the seeds were held. And we got to see, the actual seed and it was like "Wow, this the entire plant grows from this!"

Heesoon Bai argues that instead of appealing to vision-based discursive categorisation of the surroundings, a more sensuous perception arouses a participatory consciousness and nurtures an emotional relationship (148). This process was seen at the school terrace farm. It encouraged students to taste and discuss other locally grown edible plants like *shepu* (dill) and *lal math* (red amaranth) that they hadn't seen or tasted earlier. Bai further describes these intimate, embodied relationships as a process of animating the world, thereby building reciprocity and respect into relationships (as opposed to transactional interactions). Students gained an implicit notion of interdependence as they harvested the fruits of the plant they had sowed themselves a few months before. They began identifying plants based on sensory interactions, such as 'waxy leaves', 'thick leaves', 'minty taste', 'sour taste', 'sharp leaves' and so on.

The experiences were sometimes even unpleasant and unexpected, though students seemed to take it in their stride as an informative experience. Here, a student describes the sharp edge of lemongrass leaves: 'I was not very very familiar with this lemon grass. Ya I knew it is used for some tea and all but just last three and four classes back, I understood that it can cut skin also because its leaves are so sharp, I experienced it! [laughter]'.

Using the body as an 'organising core of experience' (Shusterman 51) accentuates the immediacy of experience along with a growing sensitivity to anticipated changes in the surroundings. The continuously evolving landscape of the terrace—through the growth of plants—turned into a motivation for students to explore the surroundings in a somatically grounded fashion. As a student later commented, 'Because even in gardens you see so many types of plants, but to me they were just all green, just green, a patch of green. But now I can actually like sort of at least remotely recognise that this plant is this, that plant is that and all those things.'

A 'patch of green' gaining unique features, arising from a homogenous backdrop, forms the basis for further engagement with and understanding of one's environment. Iared et. al (195) assert that eco/soma/aesthetic perceptions stimulate ontologically rich ways of relating to nature, which otherwise remain untapped or unacknowledged in discursive modes of knowledge acquisition. For instance, interactions with the growing *ambadi* (Indian roselle) plant tuned students' attention to factors promoting the health of the plant, signs of it being diseased, time of flowering, saving seeds for next season, and other related activities of plant care.

The intra-group dynamics and peer motivation was interesting to observe because the students had various levels of motivation towards participating in the project. Students who were less inclined initially found themselves participating in activities to be part of the peer group rather than the activity itself. For instance, many students were initially repulsed by the organic matter kept for composting because of general associations with it being 'waste', 'dirty' and 'yucky', but began shedding their inhibitions after seeing their friends handle it and the saplings grow in the compost. In subsequent sessions, they noticed that the compost, once prepared, had a sweet smell. Later, they began taking active interest in compost preparation and would often smell it, feel its texture, and poke around to look for earthworms (the presence of which would generate a lot of excitement). The regular activity of handling soil drew students' attention towards its texture and form, eventually leading them to observe it more closely: 'So now, wherever I go I don't decide just by looking at the soil ... If we see from above it is wet, and actually when we put our fingers we can feel it is pointy, lumpy and dry inside ... So I feel the soil now ...'.

Given that they had started out with a bare space, the emerging life-forms and relationships initiated more actions to encourage further growth. Such engrossed participation prompted a student to remark,

> We never even touched plants this way earlier ... I mean we play on the grass, but not this way. To take care ... this time we learnt how to grow the plant, otherwise it is said that just drop a seed and the plant will grow ... the book says that ... but now I think the book is very fake, because the book only says what the author can see, but while doing it we see many different things.

Bonnett ("Schools as Places of Unselving" 30) critiques formal education pedagogies for its emphasis on abstract textbook knowledge by describing schools as places of 'unselving', wherein particular histories and connections with the community, land and local context tend to get marginalised. The student's perception regarding the fakeness of the book is indicative of the gap between experience and information.

INSTANCES OF 'ENCHANTMENT'

At the farm, the plants were not objects of scrutiny; rather through their growth and other changes, the plants became active participants in expanding the students' relationship with their surroundings. This expansion in turn allowed students to attend to wider experiences and develop greater sensitivity towards the farm space. Termed here as moments of 'enchantment', the students' heightened awareness towards the farm activities allowed the usually ignored 'background' to present itself in novel, wonder-inspiring ways. This was evident in the way they would completely immerse themselves in the experience, often losing track of time until someone or something else interrupted the interaction. There was a strong affective component to these instances and all experiences weren't necessarily pleasant. For example, in one session during the monsoon, some students were completely taken aback by the sight of small wild mushrooms that seemed to have grown overnight. They went about trying to touch them, fascinated and slightly disgusted at the same time because of its almost alien-like growth on the wet pieces of cardboard planters.

Bennett comments that 'To be enchanted is to be struck and shaken by the extraordinary that lives amid the familiar and the everyday' (4). These experiences suspend control and predictability in order to make space for awe and fascination. Bennett further argues that valuing such moments 'enhances the prospect of ethical engagement' (13).

Various episodes at the farm indicated students' increasing sensitivity towards the creatures and plants on the farm. They would rescue the 'wayward' millipedes straying too far from the soil, concerned that they might die in the heat or become a prey for the crows. They would fuss endlessly around plants that were afflicted with pests. Barren soil would be carefully covered with mulch to keep the soil 'happy and moist'. Another phenomena that captivated their attention was the fruiting of flowers with swollen ovaries and dried petals. Despite 'knowing' the process through textbooks, the actual observation turned out be a novel experience for them. As one of the students exclaimed in a moment of epiphany, 'Oh, this flower is pregnant!'. The shift in view from standard images in biology to attributing a state of life to the plant (it being 'pregnant' and in need of care, and so on) seemed to be of significance for the students in relating to the plants. The variety of seeds on the farm arrested their attention, especially when they had to sow or save seeds from the harvested fruit. They would roll the okra (*bhindi*) seeds on their palm to feel its hairy texture and had a lot of fun squishing the ripe seeds of Malabar spinach which had a rich purple colour. They would hold the numerous tiny seeds of amaranth, trying to estimate how many more plants could grow from all those seeds. The amazement is evident in the following remark by a student, 'One seed can grow into one plant and we get so many seeds from it! Hundreds of them…'. In a broader, perhaps more speculative sense, we might then turn to affect not only in ethological terms, but as a wider ecology cutting across bodies, populations and assemblages (Shaw 160).

Students came to treat seeds with a lot of care and ensured that seeds were collected carefully, often spending the entire session just storing them in packets for future use. Many students started bringing apple seeds, date seeds, mango seeds, chikoo seeds, et cetera, to the farm, excited at the possibility of the seed sprouting. Others described using seeds from their kitchen at home to try growing few plants such as green gram (*moong*), mustard (*sarson*), and fenugreek (*methi*).

EMERGENT BROADER PERSPECTIVES

The activities on the farm were gradually reflected in more general thoughts students had on the environment, many of them taking shape through direct engagements or discussions on the farm. For many students, the idea of recycling took on a new meaning as they began to look for other materials which could be used as planters. On the other hand, sorting plastic from the compost led to many discussions regarding the amount of plastic in the environment and they began questioning its use in packaging, along with alternatives. Usage of dried leaves on the farm sensitised students about the usage of dried biomass in their vicinity and they made efforts to collect the biomass. Here, a student describes her growing interest in recycling: 'Actually I started recycling more because I started caring actually about the environment and the plants that I have planted. And my society as well, we had introduced a scheme that each member or each family would plant one sort of sapling or tree in the garden. So, that as well we carried out'. Others made connections with transportation of food, waste disposal and how urban farming could address the issue. For example, a student related farming to the transport of food: 'We can use each and every part of our location from ground to terrace ... and it is very advantageous because the plants are providing oxygen, the transportation of vegetables is saved, petrol is saved, and we are even making soil from dried leaves instead of throwing them in dumping grounds ...'. Another student commented on the cycle of waste to resource:

> I mean most of the waste that is degradeable and organic was ... so the waste production would be very less ... we can tell people that they can use the organic waste in the terrace garden itself. They would also be more cautious and won't produce much waste also. Like when we dump it in the landfill, we think it is just waste and don't care ... but when they themselves would see that it is affecting their plants ... in a good way ... they would care. And not just organic waste, like even the non-degradeable waste like plastic and bottles ... they could use ...

Many students found themselves empathising with farmers, as their respect for manual labour and food grew in the process of working at the farm. Remarks from a few students illustrate the point:

> I think farmers have a very tough job. We are just planting on the terrace, but they are planting on the whole field. They work in the sun, afternoon, early morning. They have to protect the field also. In our case, it is very safe and comfortable. We should really respect the farmers, because they are feeding us ...

> They (farmers) have to work a lot. Whatever we eat is what we get from them. We are making a small farm but they do cropping on a large scale, so it must be more difficult for them. Now when we see a plant infected we feel so bad. So if they have a large population of plants being infected, it must be a loss for them at a greater scale, so we think about that also ...

Their engagement with composting and adding cow dung slurry and mulch to soil helped them appreciate the richness of soil as an entity. For example, a student remarked, 'Earlier we thought soil is just something we get in packets and plants will directly grow in it. But now, we are realising that it needs cow dung, dry leaves and many decomposition materials that improve the nutrients. This has really changed what I thought about soil'.

As the above episodes illustrate, sensory experiences can create crucial emotional anchors necessary to develop a relationship with the immediate environment. Moments of 'enchantment' can emerge in an unconstrained and undirected exploration of the site. These are unique, affective episodes that can act as precursors for a more caring attitude towards one's surroundings. Feelings of novelty, challenge and a perceived sense of purpose can act as motivational triggers to enable students to feel more responsible towards the activity (farming in this case). Feedback, in the form of the evolving terrace farm, growth of plants and harvest, motivate students to remain invested in their efforts. Additionally, social feedback from peers and teachers contributed to their evolving identities as people who 'care for their environment'.

Somaesthetic Experiences as Leading to Care-Based Interactions

Their attribution of emotional states to the creatures and plants could be argued as ways of empathising with these living beings as responsive and deserving of care. Postma and Smeyers argue for the centrality of a care-based relationship in engendering environmental sensibilities, stating that these can't be derived from abstract principles of responsibility or justice. They write, 'In our caring we express a recognition of something that fulfils us in a particular way and invites the response: "When the other's reality becomes a possibility for me, I care (Noddings 14)"' (Postma and Smeyers 405).

As Ingold writes, '. . . it (Animacy) is the dynamic, transformative potential of the entire field of relations within which beings of all kinds, more or less person-like or thing-like, continually and reciprocally bring one another into existence' ("Rethinking the Animate" 10). In developing affective tendencies for the space, the overall narrative towards the place shifts as well (from 'the school terrace' to 'our farm'), further motivating students to enact their care in multiple ways.

THE CENTRALITY OF AFFECT IN GENERATING AN ETHICS OF CARE

Students' responses indicate that affective experiences can contribute to the perception of one's connection with nature, which in turn directs subsequent beliefs and actions. More generally, empathetic responses to a situation develops from the affective responses to manifold aspects of the environment. In related work, Quigley and Lyons (250) comment that environmental education is in dire need of pedagogical interventions that constitute both emotional and intellectual experiences because emotional engagement indicates deep involvement with the learners' worldviews. Sara Ahmed uses the term 'sticky' to describe the nature of affect. She writes, 'affect is what sticks, or what sustains or preserves the connection between ideas, values and objects' (29). Building a meaningful relationship with the environment requires immersive, sensorium-based experiences, which generate the rich, moral imagination necessary to think and act in pro-environmental ways. Given the atrophied and opaque nature of socioecological relationships in cities, facilitating such rich experiences and understanding the challenges in acting upon them, is a promising and urgent area of research.

The findings from this project support recent ideas about the constitutive character of artefactual engagement, particularly in generating thoughts and emotions (Ihde and Malafouris 195; Xenakis and Arnellos 1038). In this view, cognitive capacities are not just enacted but created through interaction with physical entities in the environment. I extend this argument to include the evolution of values through material involvement. The findings from the study show that the process of using entities such as compost, seeds, et cetera, was as important as the goal (in this case, growing vegetables). As Ihde and Malafouris (196) describe, the 'withness' and 'throughness' takes precedence over 'aboutness', as one is constantly 'becoming-with' (Haraway 14) and 'through' the world. The radical part of the argument denies prior existence to independent entities, insisting instead that relationships are the defining characteristic of a phenomenon. The role of 'more-than-human'[2] encounters in constituting and shaping human thought and action, when acknowledged, challenges standard notions of agency and independence, thereby creating conceptual space for enchantment, reciprocity and humility (Affifi 162).

Conclusion

Environmental education, at its core, engages with the question of why and how to care for the natural world, of which we are a part. The chapter attempts to engage with this question through tracing the evolution of empathy or a sense of connection with specific practices of gardening and growing food in one's immediate environment. Through a detailed description of students' interaction with the materials and practices involved, the study describes ways in which sustained attention can reframe the environment in terms of relations of care. These grounded perspectives can't be taught in terms of theoretical ideas.

It is not enough to be literate about ecological problems and short-term solutions. Rather, education needs to generate actions and values that shape people's way of being in the world (Chawla, "Growing Up Green" 19). Being requires becoming, through an openness to encounters that foregrounds experience over knowledge. Episodes of enchantment, as an involved activity, strengthens temporal and spatial relationship and care for one's immediate environment gets built into the process.

Heesoon Bai (140) exhorts us to snap out of the 'spell of the discursive' that lays claims on our perception by imposing an abstract, symbolic and logical view of the world. Instead, one must be willing to participate in, be affected by and care for the relations existing within the environment. In an attempt to articulate such desired perspectives towards nature, I explored farming as a practice that allows one to embody the reciprocal relationships embedded in the health of the land, soil and living beings dependent on it. Based on my findings, I argue that community farming could be an important way to motivate people to re-establish a relational ontology of respectful coexistence.

Notes

1. A detailed description of the process can be found under the resource section at the following website link: https://purvita10.wixsite.com/urbanleaves/booklets
2. According to Affifi, 'The term acknowledges and positions humans as within, as of, something bigger than is generally apparent, as it invites us to further the incomplete though ever-necessary phenomenological project of disclosing more-than-humanness in experience.' (161)

Works Cited

Affifi, Ramsey. "More-than-humanizing the Anthropocene". *The Trumpeter: Journal of Ecosophy*, vol. 32, no. 2, 2016, pp. 155–75.

Abram, David. *The Spell of the Sensuous: Perception and Language in a More-Than-Human World*. Vintage, 1996.

Ahmed, Sara. "Beyond the Affirmative Gesture". *The Affect Theory Reader*, edited by M. Gregg et al., Duke UP, 2010, pp. 29–50.

Almeida, Sylvia, and Amy Cutter-Mackenzie. "The Historical, Present and Futureness of Environmental Education in India". *Australian Journal of Environmental Education*, vol. 27, no. 1 [Special Issue], 2011, pp. 322–29.

Alsop, Steve, and Mike Watts. "Science Education and Affect". *International Journal of Science Education*, vol. 25, no. 9, 2003, pp. 1043–47.

Ashley, Martin. "Science: An Unreliable Friend to Environmental Education?". *Environmental Education Research*, vol. 6, no. 3, 2000, pp. 269–80.

Bai, Heesoon. "Re-animating the Universe: Environmental Education and Philosophical Animism". *Fields of Green: Restorying Culture, Environment and Education*, edited by M. Mackenzie et al., Hampton Press, 2009, pp. 135–52.

Barrett, M. J. "Education for the Environment: Action Competence, Becoming, and Story". *Environmental Education Research*, vol. 12, nos. 3–4, 2006, pp. 503–11.

Bennett, Jane. *The Enchantment of Modern Life: Attachments, Crossings, and Ethics*. Princeton UP, 2001.

Bird-David, Nurit. "'Animism' Revisited: Personhood, Environment, and Relational Epistemology". *Current Anthropology*, vol. 40, no. 1, 1999, pp. 67–91.

Blair, Dorothy. "The Child in the Garden: An Evaluative Review of the Benefits of School Gardening". *The Journal of Environmental Education*, vol. 40, no. 2, 2009, pp. 15–38.

Bonnett, Michael. "Environmental Concern, Moral Education and Our Place in Nature". *Journal of Moral Education*, vol. 41, no. 3, 2012, pp. 285–300.

———. "Schools as Places of Unselving: An Educational Pathology". *Exploring Education Through Phenomenology: Diverse Approaches,* edited by Gloria Dall'Alba, Wiley-Blackwell, 2009, pp. 28–40.

———. "Sustainable Development, Environmental Education, and the Significance of Being in Place". *Curriculum Journal*, vol. 24, no. 2, 2013, pp. 250–71.

Cairns, Kate. "Connecting to Food: Cultivating Children in the School Garden". *Children's Geographies*, vol. 15, no. 3, 2017, pp. 304–18.

Capra, Fritjof. *The Turning Point: Science, Society, and the Rising Culture*. Bantam, 1983.

Chawla, Louise. "Childhood Experiences Associated with Care for the Natural World: A Theoretical Framework for Empirical Results". *Children Youth and Environments*, vol. 17, no. 4, 2007, pp. 144–70.

———. "Growing Up Green: Becoming an Agent of Care for the Natural World". *The Journal of Developmental Processes*, vol. 4, no. 1, 2009, pp. 6–23.

———, and Harry Heft. "Children's Competence and the Ecology of Communities: A Functional Approach to the Evaluation of Participation". *Journal of Environmental Psychology*, vol. 22, nos. 1–2, 2002, pp. 201–16.

Cutter-Mackenzie, Amy. "Multicultural School Gardens: Creating Engaging Garden Spaces in Learning about Language, Culture, and Environment". *Canadian Journal of Environmental Education*, vol. 14, no. 1, 2009, pp. 122–35.

Dirks, Amy E., and Kathryn Orvis. "An Evaluation of the Junior Master Gardener Program in Third Grade Classrooms". *HortTechnology*, vol. 15 no. 3, 2005, pp. 443–47.

Duncan, Michael J., et al. "The Impact of a School-Based Gardening Intervention on Intentions and Behaviour Related to Fruit and Vegetable Consumption in Children". *Journal of Health Psychology*, vol. 20, no. 6, 2015, pp. 765–73.

Dutta, Deborah, and Sanjay Chandrasekharan. "Doing to Being: Farming Actions in a Community Coalesce into Pro-Environment Motivations and Values". *Environmental Education Research*, vol. 24, no. 8, 2018, pp. 1192–1210.

Dyg, Pernille Malberg, and Karen Wistoft. "Wellbeing in School Gardens–The Case of the Gardens for Bellies Food and Environmental Education Program". *Environmental Education Research*, vol. 24, no. 8, 2018, pp. 1177–91.

Green, Monica, and Iris Duhn. "The Force of Gardening: Investigating Children's Learning in a Food Garden". *Australian Journal of Environmental Education*, vol. 31, no. 1, 2015, pp. 60–73.

Haraway, Donna J. *When Species Meet*. U of Minnesota P, 2013.

Hart, Paul. "Searching For Meaning in Children's Participation in Environmental Education". *Critical Environmental and Health Education: Research Issues and Challenges*, edited by Bjarne Bruun Jensen, Karsten Schnack and Venka Simovska, Aarhus UP, 2000, pp. 7–28.

Iared, Valéria Ghisloti, et al. "The Aesthetic Experience of Nature and Hermeneutic Phenomenology". *The Journal of Environmental Education*, vol. 47, no. 3, 2016, pp. 191–201.

Ihde, Don, and Lambros Malafouris. "Homo Faber Revisited: Postphenomenology and Material Engagement Theory". *Philosophy & Technology*, vol. 32, no. 2, 2019, pp. 195–214.

Ingold, Tim. *The Perception of the Environment: Essays on Livelihood, Dwelling and Skill*. Psychology P, 2000.

———. "Rethinking the Animate, Re-animating Thought". *Ethnos*, vol. 71, no. 1, 2006, pp. 9–20.

———. *Anthropology and/as Education*. Routledge, 2017.

Jensen, Bjarne Bruun, and Karsten Schnack. "The Action Competence Approach in Environmental Education". *Environmental Education Research*, vol. 3, no. 2, 1997, pp. 163–78.

Kayumova, Shakhnoza, et al. "From Empowerment to Response-ability: Rethinking Socio-spatial, Environmental Justice, and Nature-Culture Binaries in the Context of STEM Education". *Cultural Studies of Science Education*, vol. 14, no. 1, 2019, pp. 205–29.

Kumar, Krishna. "Agricultural Modernisation and Education: Contours of a Point of Departure". *Economic and Political Weekly*, vol. 31, nos. 35–37, 1996, pp. 2367–73.

———. "Quality of Education at the Beginning of the 21st Century: Lessons from India". *Indian Educational Review*, vol. 40, no. 1, 2005, pp. 3–28.

Macfarlane, Robert. *Landmarks*. Penguin, 2015.

Moriggi, Angela, et al. "A Care-based Approach to Transformative Change: Ethically-informed Practices, Relational Response-ability & Emotional Awareness". *Ethics, Policy & Environment*, vol. 23, no. 3, 2020, pp. 281–98.

Noddings, Nel Caring: *A Relational Approach to Ethics and Moral Education*. U of California P, 2013.

Passy, Rowena. "School Gardens: Teaching and Learning Outside the Front Door". *Education 3-13*, vol. 42, no. 1, 2014, pp. 23–38.

Payne, Phillip G. "What Next? Post-critical Materialisms in Environmental Education". *Journal of Environmental Education*, vol. 47, no. 2, 2016, pp. 169–78.

Postma, Dirk Willem, and Paul Smeyers. "Like a Swallow, Moving Forward in Circles: On the Future Dimension of Environmental Care and Education". *Journal of Moral Education*, vol. 41, no. 2, 2012, pp. 399–412.

Pyyry, Noora. "Thinking with Broken Glass: Making Pedagogical Spaces of Enchantment in the City". *Environmental Education Research*, vol. 23, no. 10, 2017, pp. 1391–1401.

Quigley, Cassie F., and Renée Lyons. "The Role of Care in Environmental Education". *Exploring Emotions, Aesthetics and Wellbeing in Science Education Research*, edited by Alberto Bellochi, et al., Cultural Studies of Science Education, vol. 13, Springer, 2017, pp. 249–67.

Ratcliffe, Michelle M., et al. "The Effects of School Garden Experiences on Middle School-aged Students' Knowledge, Attitudes, and Behaviors Associated with Vegetable Consumption". *Health Promotion Practice*, vol. 12, no. 1, 2011, pp. 36–43.

Rautio, Pauliina. "Children Who Carry Stones in Their Pockets: On Autotelic Material Practices in Everyday Life". *Children's Geographies*, vol. 11, no. 4, 2013, pp. 394–408.

Raveendran, Aswathy, and Sugra Chunawala. "Values in Science: Making Sense of Biology Doctoral Students' Critical Examination of a Deterministic Claim in a Media Article". *Science Education*, vol. 99, no. 4, 2015, pp. 669–95.

Shaw, Robert. "Bringing Deleuze and Guattari Down to Earth through Gregory Bateson: Plateaus, Rhizomes and Ecosophical Subjectivity". *Theory, Culture & Society*, vol. 32, nos. 7–8, 2015, pp. 151–71.

Shusterman, Richard. "Somaesthetics and Education: Exploring the Terrain". *Knowing Bodies, Moving Minds*, edited by Liora Bresler, *Landscapes: The Arts, Aesthetics and Education*, vol. 3, Springer, 2004, pp. 51–60.

Sinha, Sujit. "Rethinking Nai Talim: What Roles Can Universities Play?". *Rethinking Universities for Development Intermediaries, Innovation and Inclusion*, edited by Shambhu Prasad and John D'Souza, Xavier Institute of Management, 2013, pp. 96–108.

Sobel, David. "Place-based Education: Connecting Classroom and Community". *Nature and Listening*, vol. 4, no. 1, 2004, pp. 1–7.

Stevenson, Robert B. "Schooling and Environmental Education: Contradictions in Purpose and Practice". *Environmental Education Research*, vol. 13, no. 2, 2007, pp. 139–53.

Wake, Susan J. "'In the Best Interests of the Child': Juggling the Geography of Children's Gardens (Between Adult Agendas and Children's Needs)". *Children's Geographies*, vol. 6, no. 4, 2008, pp. 423–35.

Xenakis, Ioannis, and Argyris Arnellos. "Aesthetic Perception and its Minimal Content: A Naturalistic Perspective". *Frontiers in Psychology*, vol. 5, no. 1038, 2014, pp. 1–15.

FOUR

READING THE ELEPHANT

TOWARDS AFFECTUAL CONCEPTUALISATIONS OF THE WILD ASIAN ELEPHANT

Sayan Banerjee & Anindya Sinha

Introduction

Asian elephants, *Elephas maximus*—through their widespread geographical distribution, adaptation, evolution and extinction—have fascinated humans throughout history. Given their ecological function as ecosystem engineers and their iconic status within religion, history and popular culture—especially in India—the relationship between Asian elephants and humans has always been a vibrant, well documented one (Sukumar). Several disciplines across the sciences and humanities have attempted to conceptualise what an elephant is and what an elephant does: is it only an environmentally determined, automated life form or is it capable of sentience and emotion? How complex is its life and how free and wilful their decision-making processes? Asian elephants have thus been an effective agent in establishing dialogues across disciplines and in promoting various controversies within these disciplines. In this paper, we first discuss these disciplinary turns and later offer an alternative way to engage with the lives and times of elephants in a world we have increasingly come to share with them.

The scientific understanding of human–elephant relations is of fairly modern origin and has been traditionally dominated by the biological disciplines of ecology, ethology and conservation biology. These disciplines have mainly examined the elephant in isolation: documenting its socioecology, behavioural ecology, resource use, physiology or cognitive abilities while leaving aside their interactions with humans, which may have significantly shaped these aspects of their lives in recent times. Demographic charts, ethograms and time–activity budgets thus become the imperative to understand how elephants exist; the nature and extent of their responses to humans forgotten. However, given our observations that most elephant movements occur across human-dominated and anthropogenically modified landscapes, humans do occasionally appear at the periphery of these studies, albeit marginally and as a mere agent of the Anthropocene. More specifically, when some studies on human–elephant 'conflict' in India have come to the fore, they have documented certain patterns of this conflict—such as frequency and seasonality of raiding, extent of crops damaged or the fatalities incurred (Choudhury; Gubbi et al; Wilson et al.)—but have paid little attention to the perceptions and attitudes of the local human communities towards the elephants they share their lives with. This could, of course, conceivably be due partly to biologist's lack of interest in understanding the impact of humans on elephant ecology and behaviour, but also possibly due to their incomplete understanding of and ambivalence towards the research methods of the social sciences. Critiquing this paucity of attention to the human dimensions of human–elephant relations, social scientists (especially geographers and anthropologists) have made attempts to understand the complexities of human–elephant relationships from the perspective of humans who share space with elephants (Ogra "Human–Wildlife Conflict"; Jadhav and Barua; Barua "Bio-geo-graphy", "Volatile Ecologies"; Gogoi).

Ethnographies, employing observation- and interview-based tools have thus been developed to document the often severe negative impacts of human–elephant conflict on the livelihoods and the quality of life of the people who share living spaces with elephants, as well as how these impacts and the responses to them have been mediated through social structures of class, caste, gender or ethnicity (Ogra, "Human–Wildlife Conflict", "Attitudes Toward Resolution of Human–Wildlife Conflict"; Barua et al.; Banerjee). In these studies, however, the elephant has been pushed to the margin and treated as an object that has transgressed the boundaries of its human-assigned spaces unlawfully. Through such an examination of human–elephant relations from the narrow lens of disciplinary perspectives therefore, the natural and social sciences have established separate in-depth understandings of human and elephant worlds. Both these disciplines have compartmentalised the relationship into conflict and coexistence with mechanistic indicators employed to measure the extent of this relationship.

A recent subfield within geography has critiqued the prevailing approach by proposing that human and elephant systems are not independent of one another; rather the lives and landscapes over which the encounters between the two species and their subsequent interactions occur are co-constructed by humans with elephants (Locke; Barua, "Bio-geo-graphy"). This more-than-human turn in geography has called for examining the affective atmospheres of animals (Lorimer et al.), the integration of individual and collective animal subjectivities into geography (Bear) and establishing dialogues between geography and ethology (Barua and Sinha). Such studies have gained traction over the last decade and new insights on interspecies working relations and relatedness have been brought forward (Locke; Barua, "Bio-geo-graphy"; Lainé; Münster; Keil), with the elephant being at the forefront of the examined species. The more-than-human geography project largely being confined to the scholars from the global North has also risked reproducing the colonial hegemony of knowledge production by silencing Indigenous epistemologies (Sundberg). The people who actually co-inhabit and co-construct their lives with elephants, thus, often remain voiceless in this more-than-human scholarship.

Our paper attempts to foreground an alternative account of engaging with human–elephant relations—locating itself within the realm of the more-than-human tradition with a focus on affect and perceptions and also within decolonial thinking—by engaging with the situated knowledge and testimonies of the local human communities who share space with elephants every day.

Locating the Study

The Udalguri district in the state of Assam, situated on the Indo–Bhutanese border in north-eastern India, experiences some of the most difficult aspects of human–elephant interactions anywhere in the country. The Dhansiri Forest Division in this district is particularly vulnerable. This landscape, part of the larger Chirang–Ripu Elephant Reserve is a mosaic of agricultural settlements, tea gardens and reserve forests. The Khalingduar Reserve Forest and its adjoining areas in particular, witness substantial movements of about 150 Asian elephants (*Elephas maximus*)—both herds and individuals—for almost six to seven months of the year (Assam Forest Department). Most of this movement occurs through non-forested areas and human–elephant encounters thus become an on-ground reality every day. Negative human–elephant incidents have increased over the years. This has been ascribed to the depletion and degradation of elephant habitats in the Khalingduar Reserve Forest increasing long range migration of elephants and the growing anthropogenic land use changes around these forests. Consequently, 54 elephants and 121 people have died, 8333 houses damaged or destroyed and approximately 1400 hectares of cropland depredated from 2007 to 2016, according to a Divisional Forest Officer in the Dhansiri Forest Division.

The local human communities in this predominantly rural area ethnically consist of Bodos, Nepalis, Adivasis and Bengalis, with agriculture being their major source of livelihood. The tea plantations which were set up in late nineteenth century, however, continue to play a major role in the local economy. The crop fields in the region appear to provision the migrating elephants with there being a significant increase in house or crop damage and in human or elephant mortality from June to November every year, coinciding with the agricultural calendar (Figure 4.1). Apart from death or injury to both humans and elephants there are several other impacts that have been documented, albeit only within humans. The mobility of the affected people is restricted and they typically experience increased workloads and expenditure to overcome their personal and livelihood losses. These, in turn, lead to often silent compromises in their mental health. Such indirect impacts of human–elephant encounters are often long-term but remain unacknowledged by conventional, short-term coping strategies such as crop guarding or monetary compensation which are usually initiated by the local Forest Department.

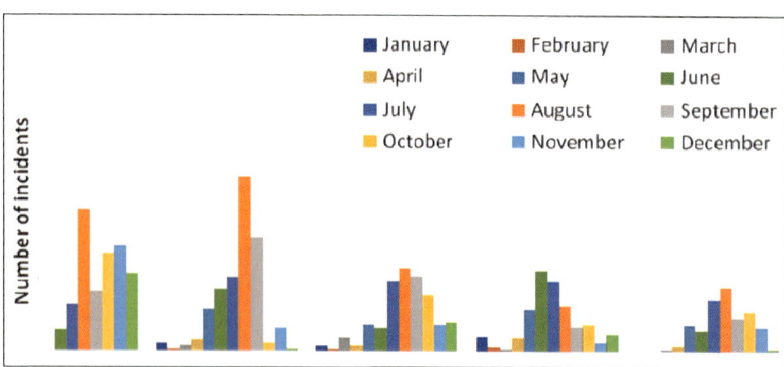

Figure 4.1: Temporal variation in incidents of negative human–elephant interactions in the Udalguri district of Assam, from 2012 to 2016

One of us (SB) conducted a field survey between 2016 and 2017—covering twelve villages falling within the jurisdiction of Dhansiri Forest Division—to document certain aspects of the lives of people who encountered wild elephants in the everyday. These villages reported relatively higher incidences of negative human–elephant interactions as compared to other villages in the region and were situated in the vicinity of either the Khalingduar Reserve Forest or its neighbouring tea estates. In-depth interviews related to the nature and patterns of human–elephant encounters and interactions; impacts of these interactions; coping strategies; experience with elephants, as well as one's attitudes towards and perspectives of elephants were conducted with 65 respondents between 18 and 76 years of age, 30 of whom were women.

What emerged strongly in our study across this rural landscape were gendered patterns of socioeconomic roles and responsibilities—both domestic and productive—and their subsequent evocation of space utilisation (Table 4.1), often leading to narratives that centred around the specific sharing of spaces between women and elephants. As the women who we interviewed lived and worked in the forests, along the riverbanks and in tea plantations through which elephants also frequently moved, fed or rested, they encountered elephants more than their men did. The men interacted actively with elephants mostly in the agricultural fields while guarding crops or when they intoxicatedly walked into a herd, usually losing their lives in relatively greater numbers than the women.

Table 4.1: Gendered division of labour and utilisation of space, Udalguri district of Assam, 2016–2017

	Gendered division of labour	*Gendered utilisation of space*
Women	• Sowing • Harvesting • Firewood collection • Drinking water collection • Household work	• Forests • River and other water sources • Tea estates • Agricultural fields • Households
Men	• Land preparation for agriculture • Livestock grazing • Crop guarding • Daily wage-based labour	• Agricultural fields • Grazing lands • Marketplaces • Other villages/towns

PERCEPTION, EMPATHY, ANTHROPOMORPHISM

Humans and nonhumans alike make sense of their external environment, including any living or non-living entity, by receiving sensory information about them. Such information is organised, processed and interpreted against the background of their evolutionarily acquired biological properties and their developmentally acquired individual experience, thereby building up a robust perception of the entity. The people in our study landscape clearly co-constructed their lives with elephants on an everyday basis and appeared to organise their auditory and visual experiences of these elephants to form certain conceptualisations, with associated knowledges, of what an elephant is.

Myths related to elephants and their different behaviours, information on the nature of human–elephant interactions from other places—near and far—and elephant vocalisations perceived during encounters allow for the development of an auditory repertoire within the perceptual information that people acquire about elephants. Our interviews revealed that the active evaluation of these auditory resources made certain people consider elephants to be godlike creatures but a risky species to share space with. Close as well as distant physical encounters with elephants in different spaces—such as forests, crop fields, plantations or around settlements—on the other hand, provided our subjects with visual cues through which they were able to form visual representations of what elephants can achieve and the nature of their behavioural responses to human actions, particularly within shared spaces. It also became apparent to us that direct visual information was instrumental in making people form more nuanced perceptions and construct lively visualisations of the elephants (see **Testimonies** below).

Testimonies (translated from Assamese to English)

'After all these years, I have seen elephants from so close that all my fear is gone. They have become *goru* [cow] for me. My little kid also does not get afraid.'

—Ravi Tanti, male, 45

'Elephants too have to feed themselves. If we get hungry, we search for food. The elephants have such a large stomach. They have to feed well.'

—Radhika Rai, female, 37

'Elephant lives are not different from ours. We have similar requirements. They come here only for food, as we go to Gujarat or Mumbai to work. They do not have anything to eat in the forest.'

—Keshab Pradhan, male, 41

'When we need firewood, we make a group of three to four women and go to the *bagan* [forested patch]. If we see elephants roaming at a distance, we first pray that we have not committed any sin, so we should not be harmed. We then enter and collect whatever broken twigs are there on the ground. We can see the elephants constantly looking at us and giving us blessings. They are very intelligent beings.'

—Meera Pradhan, female, 56

'Elephants know who have committed *paap* [sin]. They do not spare them. They come to the sinner's place and break the house down. I have not done any sin, so it has never broken down my house.'

—Anima Boro, female, 61

'It almost acts like a police. Due to the elephants' presence in the night, nobody dares to move, not even thieves, dacoits or NDFB [an insurgent group]. In other places, these terrorists come in the night and create ruckus. But due to the elephants, they cannot come here. Yes, our movement is restricted, but that is fine with us.'

—Mukto Saharia, male, 38

'Every elephant is not bad. Only two to three out of a hundred are bad and come only to damage houses. These are like the local *dada* [local goons]. Even the herds do not accept them. They try to enter the herd, but the herd members push them out. They then follow the herd at some distance. These are the elephants that kill people.'

—Kaneshwar Medhi, male, 53

'Earlier the elephant used to be fearful of us. It only came to our field, but never damaged anything. If we even just clapped our hands, they would run away. Now they have changed. Even though we have fire in our hands, they come running straight towards the fire. It is only when more people are gathered and burn some noisy crackers that they move from here.'

—Lakhiram Boro, male, 78

'Elephants can move anywhere they want. But why do they come here? It is because their lifestyle has changed. They have seen humans and learnt about their food.'

—Krishna Thapa, male, 51

Overall, the testimonies revealed a gendered view of the elephants and their life-history strategies, as well as subtly gendered interpretations of their behavioural responses towards humans. Going even beyond was a gendering of the elephants themselves. Many respondents thus equated certain lone bull elephants—whether tuskers or *makhnas* (widely used vernacular term for tuskless male elephants)—with 'reckless youth' or 'Salman Khan' (a noted actor in the Hindi film industry known for his daredevil stunts), or even a 'terrorist' due to their aggressive attitude and abilities to take risks. The elephant herds consisting of adult and subadult females and their dependent calves and juveniles, in contrast, were often compared to calm and docile livestock due to their relaxed body postures and tendency to retreat, instead of displaying aggression, when disturbed.

There were also reflections on the change in behavioural profiles of the elephants over time. Some of the respondents claimed that the elephants had become more aggressive than before, as they were now more 'habituated' to living close to humans and had 'experienced' human aggression. The everyday observations of elephants also produced empathy amongst men and women alike. They often projected the underlying rationale of certain behaviours displayed by elephants as akin to the miseries in their own life, especially those magnified by their poverty and reduced opportunities. The elephants were thus anthropomorphised into beings, whose actions and responses to humans could be justified and the impacts of living close to them accepted as being almost inevitable. Such an anthropomorphic framework not only led to the perception of elephants as fellow compatriots in a shared world, but also allowed people to reflect on their own lives, destined to be intertwined with those of elephants across time.

VERNACULAR ETHOLOGISTS AND THEIR ETHOLOGIES

Local human communities that share space with wild elephants in the everyday possibly find it imperative to have certain active conceptual representations of elephants as integral components of their own lives. Anthropomorphism then becomes an important tool that enables far richer, local understandings of elephant decision-making processes and life-history strategies, in addition to providing a bridge through which elephant-like and human-like characteristics can be fused to comprise an apparently co-constructed, existential reality. We suggest that the perceived commonalities in their shared lives arise, at least in part, from their sense of shared marginalities: a world in which both elephants and humans are driven to a precarious existence by forces beyond their control, the former by the large-scale loss of habitats—homes for the elephant—and the latter by unending intergenerational cycles of poverty and deprivation. The intensity of their cohabitation appears to have compelled people to employ all their senses to understand and document various aspects of the lives of wild elephants and assimilate them into their own pathways of living. What is striking is that, rather than being a monolithic entity, the elephant seems to have taken on a rather amorphous form in such a worldview: god, beast, terrorist, police, neighbour, cow or mother, perhaps all at the same instance. Moreover, the personal nature of the interactions that each human experiences leads to a clear conception that the elephant is not a singular species. The individual elephant emerges in all its nuances. Each elephant is now a unique personality, temperamental in its own ways. What is most remarkable is that these embodied conceptualisations—the local understandings of the elephant—diverge from the increasingly impotent visualisation of the elephant as a collective—a mere species—so popular in mainstream zoology-dominated thinking (Sinha; Srinivasaiah et al.). Our biological thinking, driven largely by Western philosophical traditions, has typically treated such anthropomorphic worldviews as being patently unscientific and disdainfully anecdotal. The time has come to make amends, to recognise the powerful existence of vernacular ecologies served by relatively deep affective and ethological understandings, which are in turn generated by the bodily experiences of individual humans living their lives in close intimacy with individual elephants: with other more-than-humans as well.

READING AN ELEPHANT

How then is an elephant understood? It is evident that both short and long-range observations and sensations—some direct, some indirect, some in shared spaces, some at distant locations—shape vernacular human perceptions of elephants. Close encounters in the crop field may be sudden, marred by confusion due to the limited visibility of elephants in the darkness. Noises created by elephants—some powerfully semiotic, some perhaps meaningless though diagnostic—and bodily movements trapped in small torchlights constitute perhaps the only source to quickly comprehend the intent of the other. Encounters at an intermediate distance, on a road or in a plantation, provide for better observations; some affectively shaped and registered. When more distant, in a neighbouring village or across the hill, people and elephants may not interact with one another at all but their presence is felt; their responses heard of; opinions formed by the local gossip; their intentionalities, however, clearly drawn. Over time, across generations, the human and the elephant become relatives: intimacies are born. Through such intergenerational observations, we also encounter people's concerns about the changing, still unfamiliar behavioural patterns of the familial elephants. Some descriptions, fully engaged with every turn or every movement of an elephant in this shared landscape, appear to be akin to reading and engaging with local literature: rich in its knowledge but yet to be revealed fully.

All our respondents were able to recall the auditory and visual memories of elephants they had associated with or experienced, directly or indirectly, and were able to express themselves emphatically in the available vocabularies and language. The limitation of this exercise was, of course, that the immediate affects, emotions and embodied responses elicited during the actual encounters with elephants remained unexplored. Such affects, though experienced intensely, are often inexpressible and often so momentary that they fail to be captured in language post the encounter. Whether they are even encapsulated in our memory also remains an open question. It is, however, crucial to note that such affects nevertheless contribute to the ultimate repertoire of concepts that define an elephant to the experiencer. Continuing with our analogy of literary experience, the conceptual unravelling of the elephant through interview-based methods then appears to be a mere 'review of the elephant', akin to a book review, failing to engage with an individual's more direct ways of reading the elephant. Our methods thus remain inadequate to unpack the entire experience of being with an elephant; they only provide a glimpse into the depths of prevailing human–elephant relations; a particular view of their co-constructed world. We therefore suggest that novel, richer, ethnographic methodologies be developed to comprehensively document the human–elephant ethnologies developed by people across generations.

To return to our comparative account of different methodologies, an important aspect of our understanding lies in being able to clearly differentiate between various kinds of encounters, usually generated by elephants but interpreted by humans, for we do not yet completely understand the bodily and mental experience generated within the nonhuman individual herself. Encounters in human–wild elephant interaction research are only considered as such when they have a tangible output including, for example, the number of deaths or injuries caused, or the statistics related to the assets or resources that are damaged. We, however, suggest that the nature of each encounter is unique, shaped by the time and space over which the encounter takes place and the socioeconomic location and experiential history of the individual participating in the encounter. The embodiment of each encounter is extremely variable and could range from downright physical combat with aggressive bodily responses from either side to more ritualistic performances with distant eye contact or none at all and with minimal or no physical response from any of the actors. It then becomes imperative for researchers to evaluate and ascertain how people's perceptions and conceptualisations of elephants can potentially be understood on the basis of the different encounters that they have experienced. The testimonies that spell out a rather amorphous anthropomorphism of elephants themselves provide evidence that the experiencers treat individual encounters to be different and not merely as simply classifiable interactions. In order to examine the affective dimensions that human–elephant encounters typically produce, it becomes crucial to document the nature of the actual encounters and the resulting embodied responses from both elephants and humans. Although non-participant observation-based research designs could potentially be employed in these contexts, we postulate that certain kinds of participant observations within which the observer becomes (admittedly) empathically involved with the participants of the interactions, could generate novel affective methodologies best suited for the purpose (Sinha et al.; Ramakrishna et al.).

ALTERNATE UNDERSTANDINGS

The relationalities espoused by the local communities while interacting with wild elephants in the everyday lead to internalised ways of being and living with elephants, thus dismantling the strict compartmentalisations and definitions of conflict and coexistence put forward by conventional management thinking. The human–elephant relationship in our study area represented a rather pulsating equilibrium, hovering within a repertoire of affect-, emotion- and action-driven perceptions. The specific expressive used for the elephants by the local people also situates the elephants in a particular conceptual domain, likely to be geographically unique. With more empirical studies of this nature, however, the notion of 'relationship' can itself perhaps be changed to 'relationships', thereby spatially and temporally grounding the specific, albeit unique, nature of the encounters and interactions observed in our study to a specific set of humans and elephants. On the basis of these emerging specificities, therefore, we would like to question and challenge the generalised conflict mitigation models almost invariably proposed by the mainstream human–elephant interaction management regimes. These relations are, of course, further complicated, at least in the human dimension, by the place based social realities of class/caste/gender/ethnicity mediated individual and group experiences of living close to wild elephants.

Wildlife management and conservation models in India typically propound voluntary or forced behavioural management of local human communities by setting up resource-use regimes and spatial boundaries that propose to keep out the wildlife, often to their detriment. It is often assumed that 'conservation education and outreach' modules, prepared by external 'experts', with a rather mainstream understanding of elephants, have the potential to promote 'coexistence' with wildlife. Such interventions, however, hardly take into account place-based human perceptions of elephants and hence, overlook or even deny the notion of a relational coexistence as understood and experienced by the local people. These conservation models typically maintain an unequal power structure, whereby certain traditional knowledge systems and considerations are invariably marginalised as anecdotes, thus raising doubts even about their authenticity let alone their practical importance. Mainstreaming alternative understandings of elephants by engaging with intensely localised place and time based anthropomorphisms can and will dismantle these power structures and serve to bring the perspectives and opinions of the local people to the forefront.

The Way Forward: Readings Beyond the Human

A final point concerns our observation that affectual conceptualisations of human–elephant relations remain only partially understood if we only consider the human dimensions of these relationships. What about affect in elephant lives? Such a consideration confronts us with various difficulties, both conceptual and methodological. Although affect studies have long presented the importance of animal subjectivities and sentience, empirical studies related to documenting affect in animal lives have been scarce (but see Sinha et al.). Given our failure to interrogate nonhumans about their perceptions, especially towards humans with whom their relations appear to be increasingly troubled in the Anthropocene, participant observation-based methods appear to be the only way forward to understand affect in non-verbalising cultures. Traditional ethological studies are indeed observation oriented but are highly quantitative and structured with predefined times and spaces, as well as individuals, randomly chosen for observation. There have been recent attempts to understand human–animal relatedness through labour and kinship, with ethnography being the primary method of choice (Barua, "Bio-geo-graphy", "Volatile Ecologies"; Münster). Although ethnography has principally been used to study human behaviour, cultures and perspectives, can we bring the reflexivity, flexibility, depth and unstructuredness typically associated with it to examine the more-than-human as well? Can we do rich, more complete ethnographies of elephant lives?

Works Cited

Elephants in Assam. Assam Forest Department, 2010.

Banerjee, Sayan. *'En-gendering' Human-Elephant Conflict (HEC): The Case of Udalguri District of Assam*. 2017. Tata Institute of Social Sciences, Master's Thesis.

Barua, Maan. "Volatile Ecologies: Towards a Material Politics of Human-Animal Relations". *Environment and Planning A*, vol. 46, no. 6, 2014, pp. 1462–78.

———. "Bio-geo-graphy: Landscape, Dwelling, and the Political Ecology of Human-Elephant Relations". *Environment and Planning D: Society and Space*, vol. 32, no. 5, 2014, pp. 915–34.

———, and Anindya Sinha. "Animating The Urban: An Ethological and Geographical Conversation". *Social and Cultural Geography*, vol. 20, no. 8, 2019, pp. 1160–180.

———, et al. "The Hidden Dimensions of Human–Wildlife Conflict: Health Impacts, Opportunity and Transaction Costs". *Biological Conservation*, vol. 157, 2013, pp. 309–16.

Bear, Christopher. "Being Angelica? Exploring Individual Animal Geographies". *Area*, vol. 43, no. 3, 2011, pp. 297–304.

Choudhury, Anwaruddin. "Human-Elephant Conflicts in Northeast India". *Human Dimensions of Wildlife*, vol. 9, no. 4, 2004, pp. 261–70.

Gogoi, Mayuri. "Emotional Coping among Communities Affected by Wildlife–Caused Damage in North-East India: Opportunities for Building Tolerance and Improving Conservation Outcomes". *Oryx*, vol. 52, no. 2, 2018, pp. 214–19.

Gubbi, Sanjay, et al. "An Elephantine Challenge: Human-Elephant Conflict Distribution in the Largest Asian Elephant Population, Southern India". *Biodiversity and Conservation*, vol. 23, no. 3, 2014, pp. 633–47.

Jadhav, Sushrut, and Maan Barua. "The Elephant Vanishes: Impact of Human-Elephant Conflict on People's Wellbeing". *Health and Place*, vol. 18, no. 6, 2012, pp. 1356–365.

Keil, Paul G. "Uncertain Encounters With Wild Elephants in Assam, Northeast India". *Journal of Religious and Political Practice*, vol. 3, no. 3, 2017, pp. 196–211.

Lainé, Nicolas. "Conduct and Collaboration in Human-Elephant Working Communities of Northeast India". *Conflict, Negotiation, and Coexistence: Rethinking Human-Elephant Relations in South Asia*, edited by Piers Locke and Jane Buckingham, Oxford UP, 2016, pp. 180–205.

Locke, Piers. "Explorations in Ethnoelephantology: Social, Historical, and Ecological Intersections between Asian Elephants and Humans". *Environment and Society*, vol. 4, no. 1, 2013, pp. 79–97.

Lorimer, Jamie, et al. "Animals' Atmospheres". *Progress in Human Geography*, vol. 43, no. 1, 2019, pp. 26–45.

Münster, Ursula. "Working For The Forest: The Ambivalent Intimacies of Human-Elephant Collaboration in South Indian Wildlife Conservation". *Ethnos*, vol. 81 no. 3, 2016, pp. 425–47.

Ogra, Monica V. "Human–Wildlife Conflict and Gender in Protected Area Borderlands: A Case Study of Costs, Perceptions, and Vulnerabilities from Uttarakhand (Uttaranchal), India". *Geoforum*, vol. 39, no. 3, 2008, pp. 1408–422.

———. "Attitudes toward Resolution of Human–Wildlife Conflict among Forest-Dependent Agriculturalists near Rajaji National Park, India". *Human Ecology*, vol. 37, no. 2, 2009, pp. 161–77.

Ramakrishna, Ishika, et al. "Ethos, Pathos, Logos: Affective and Emotive Ethologies of Primate Lives." *Ecological Entanglements: Affect, Embodiment and Ethics of Care*, edited by Ambika Aiyadurai, Nishaant Choksi and Arka Chattopadhyay, Orient BlackSwan, 2022, pp. 15–40.

Sinha, Anindya. "Not In Their Genes: Phenotypic Flexibility, Behavioural Traditions and Cultural Evolution in Wild Bonnet Macaques". *Journal of Biosciences*, vol. 30, no. 1, 2005, pp. 51–64.

———, et al. "Affective Ethnographies of Animal Lives". *A Research Agenda for Animal Geographies*, edited by Alice Hovorka, Sandra McCubbin and Lauren Van Patter. Edward Elgar Publishing, 2021, pp. 129–46.

Srinivasaiah, Nishant M., et al. "Usual Populations, Unusual Individuals: Insights into the Behavior and Management of Asian Elephants in Fragmented Landscapes" *PLOS One*, vol. 7, no. 8, 2012, e42571.

Sukumar, Raman. *The Living Elephants: Evolutionary Ecology, Behaviour, and Conservation*. Oxford UP, 2003.

Sundberg, Juanita. "Decolonizing Posthumanist Geographies". *Cultural Geographies*, vol. 21, no. 1, 2014, pp. 33–47.

Wilson, Scott, et al. "Understanding Spatial and Temporal Patterns of Human-Elephant Conflict in Assam, India". *Oryx*, vol. 49, no. 1, 2015, pp. 140–49.

FIVE

BLINDNESS AND CANINE HEROICISATION

INTERDEPENDENCE IN KUUSISTO'S *HAVE DOG, WILL TRAVEL*

Krishna Kumar S

The companionship between a blind person and his guide dog epitomises the possibility of an intimate, non-hierarchical interconnection between humans and other species. In navigating the physical world together, the dog and its human companion become 'extensions of each other' based on an attunement to their distinct, yet relatable, subjectivities (Michalko 5). They gear themselves to one another while facing the world as a harnessed dyad, bound by a sense of 'affective relationality' (Nishida 101).[1] Both contribute equally to the efficacy of this companionship in that, while the dog aids her blind owner's safe passage, the latter bestows their unreserved trust on his canine companion, thereby enabling it to operate with confidence. However, the sighted world refuses to acknowledge this relationship as interdependent, choosing instead to heroicise the dog as having absolute control over the mobility of an otherwise helpless blind person. The heroicisation results from, and further reinforces, the negative signification of blindness as a state of 'lack' and of abject 'dependence' (Esmail 19). Although heroicising the canine seems to be an exercise in inverting the 'human/non-human animal hierarchy', it is problematic since it reinstates the archetypal alignment between sight and knowledge/ ability on one hand and blindness and ineptitude on the other (19). Given this conundrum, it is essential to inquire into the possibility of a middle ground between anthropocentrism and an exclusive heroicisation of the canine. The chapter argues that a foregrounding of the dynamics of trust and interdependence underpinning the canine–human relationship counters exclusivisms of either kind mentioned above. It does so by examining Stephen Kuusisto's *Have Dog, Will Travel* (2018), a memoir that presents the author's narrative attempt at retaining the integrity of his blind self while affirming the importance of his canine companion to the assertive re-presentation of his disabled identity.

The memoir chosen for the study traces Kuusisto's canine-induced transformation from a life of passing as sighted to one of assertion in blindness. It proclaims his adoption of a guide dog as a catalyst that helps him overcome the compulsion to mimic normalcy and come out as blind. It shows that in response to the erasure of his presence brought about by the heroicisation of the dog, he is forced to reassess his transformation. To begin with, he revisits his pre-dog past to reclaim its positive bearing upon his present self. He discovers that impulses developed during his days of passing continue to linger as residual presences and subliminally empower his canine-assisted orientation to the world. This leads him to frame his past and the canine-inflected present as fluid temporalities that mesh into each other to produce a complex narrative of blindness.

Second, he emphasises trust, rather than power, as a pivot that anchors his relationship with the canine other. Drawing on the notion of 'gift of trust' posited by Hannah Macpherson (132) and the concept of 'inter-corporeal generosity' discussed by Stevenson Andrew (1165), the chapter reads the reciprocation of trust between Kuusisto and his guide dog Corky as an essential affective prerequisite to the efficacy of the companionship. In doing so, it reinstates the importance of understanding this interspecies relationship through a framework of interdependence, rather than one of exclusive heroicisation of the dog to the estrangement of its human partner. It teases out the aesthetic of interdependence channelled through agentic dispersal and mutual care acts.

STEPPING OUT OF THE 'SHELL' INTO THE OPEN

The inability to project one's 'actual' self for fear of being branded as deviant from the 'normal' forms the first of two central crises in Kuusisto's life (Goffman 134).[2] The epigraph to the memoir—an axiom by a Buddhist abbot called Chögyam Trungpa—captures the issue: 'The ultimate definition of bravery is not being afraid of who you are' (Kuusisto ix). Taught to disavow his impairment from childhood onwards, Kuusisto acquires a split sense of self and inheres anxiety to be who he is. The disavowal compels him to pass as sighted even as his 'blind soul' remains 'quiet in its shell,' and urges him to develop folk techniques that would enable him to sustain the artifice of normalcy (33).

I discuss two techniques in particular to show the extent to which he has internalised the necessity to maintain a 'virtual' identity of sightedness (Goffman 134). They include (a) developing a mnemonic familiarity with the immediate terrain; and (b) navigating through guesswork.

The persistent anxiety regarding the exposure of his blindness makes Kuusisto commit the entire dimensions of a space to memory in order to maintain the impression of effortless mobility. He prefers small, secure places that would help him weave a personalised orbit around which he could effectively pass without accident (Kuusisto 13). He says that he 'knew every inch of the (college) campus' where he studied and admits that at such a small school it was easy to pretend that he could see (13). When he wishes to move to a larger landscape—for instance, to Iowa City for further studies—he flies there earlier than needed to acclimatise himself with its geographical coordinates (14). He writes of moving slowly through the city streets, taking out a telescope to look at the street signs when he thought no one was watching (14). This indicates that a larger terrain with a greater sense of unpredictability tends to pose a graver challenge to the sighted performance of a passing blind subject. It confirms that effective passing depends on both the demography as well as physicality of a space. Alongside this almost pathological urgency to contain a terrain within his memory, Kuusisto also tends to navigate based on dare and assumption. He defines such a method as follows: 'faking sight is like being illiterate—you pretend to [have] competence but live by guesswork' (21). That is, the one who pretends must make it seem natural and hold it together even if it is fraught with peril. For instance, during a college excursion to Santorini, he goes for a bike ride by dint of audacious guesswork using his fractional sight that is always already suspect (11). Keeping the 'bright red windbreaker' of a classmate called Timothy as a compass, he rides the bike like the others, describing the experience as being in a bullfight (11). The perceptual metamorphosis of a windbreaker becoming 'a rectangle of red' and then 'a red flag' hints at the world sliding swiftly away from the grasp of the visually-impaired subject and shows his difficulty in processing the discrepancy between 'reality' and his subjective perception of it. Ironically, the effort involved in the act also reveals that passing based on guesswork requires a certain amount of competence and skill on the part of the pretending subject.[3]

The concealment of one's corporeal difference depends on the world's willingness to overlook the performativity of passing. Kuusisto resorts to a gaming analogy to explain his inevitable reliance on the world for the approval of his sighted 'charade' (14). He writes, 'While pretending to see, reality was my opposing driver. . . . Would "the real" step aside for my blind race?' (10). This is effective only as far as the 'real' shows benevolence towards the subject; the moment it fails to play its part, it may even endanger the very existence of the latter. He realises this in Ithaca where, 'following people who looked like shadows' as is his wont, he is almost killed by a car (17). The near death proves the volatile nature of his lopsided pact with the world and shows him that he can no longer pass without putting his very mortality at risk. It makes him realise that he needs a 'partner' who only cares about his safety (17). In other words, he wishes for a partner who will be an active agent of care and decides to adopt a guide dog. He retrospectively remarks on the decision that little did he know he was going to 'enter' his guide-dog's story (18). His declaration that he 'was about to enter' his guide dog's story rather than vice versa places the dog at the centre of this narrative of companionship and affirms that it is as much a canine event as a human one.

The canine arrival marks a moment of 'bestia ex machina' for the author whose 'blind part' of the self was 'starving' at the time (Kreilkamp 39; Kuusisto 9). Its regenerative potential lies in the dog's non-judgmental attitude towards her human partner (6). For the dog, her companion's past is largely irrelevant as she incorporates him in her story anew. For her, he remains only as a 'present-tense man'; one who needs fulltime assistance in safe navigation of the world (9). This dog-invoked emphasis on presentness urges him to move on from his troubled past. Hence, he bids adieu to his pre-dog life, 'goodbye blind éclair, the one with the shame frosting . . .' (12). The declaration highlights the importance of the ritual of self-cleansing when entering the lifeworld of another being. Rendered in the form of a farewell note—rather than reflectively woven into the narration—it underpins Kuusisto's realisation that he needs to consign his internalised inhibitions to the past before he can fully own the canine-inflected present. Accordingly, he lays the feelings of shame and diffidence—accrued both consciously and unconsciously—to rest so that he can approach his life afresh with the company of a guide dog.

Kuusisto's coming out as blind can be understood in both senses of the act: 'self-acceptance' as well as 'the process of revealing or explaining one's disability to others' (Samuels 348). The canine association, to start with, makes him realise that his corporeal identity is not as complicated as he imagined (Kuusisto 132). This is the voice of a person whose subjectivity had been entangled in complex knots in his interaction with an alienating world, who ultimately realises that it is his relationship with a guide dog that leads to a simplified understanding of self and, consequently, acceptance of himself as he is. Second, the canine copresence draws his 'blind soul' out of its 'shell' and enable him to represent it unapologetically in the visible world (33). Being teamed up with Corky and having to no longer to worry about the precarity of his presence, he confronts the world head on and challenges it to respond to his reconfigured orientation. He nonchalantly declares that the world could figure itself out. He also relishes the simplicity of representing the self in all its ordinariness: 'You're just you' (141). This signals a sense of compatibility between the felt and projected identities of the subject, one that eluded him in all his pre-dog existence.

The memoir also shows that one of the most significant modes through which Kuusisto and Corky forge 'affective relationality' is intersubjective sensitivity (Nishida 101). It unfolds as a somatic praxis whereby man and dog remain ever alert to one another's 'behavioural repertoire' while coursing through the space in seamless tandem (Andrew 1163). The reciprocity of affects that propels this dyad is apparent in the manner in which Kuusisto talks about the deep affection and discernment that exists between him and Corky. This points to an affective interconnection based on an admixture of mutual attentiveness and an emotional bond whose depth can be gauged from the permeability of sensations and stimuli across the human and the canine: 'I was calm because Corky was calm. Dogs reflect us and our moods' (108). The passage shows that affects refract across beings in a companionship even if they belong to different species, which indicates the possibility of forging an intimate interconnection between humans and nonhuman animals at a corporeal level. The fact that it begins with Kuusisto ascribing his tranquillity to his dog and ends with the latter's calmness being tied to the former, reflects the refractability of internal states across species. This temperamental synchrony underscores the affective foundation of the interdependent relationship between the blind person and his service dog.

THE CRISIS OF CANINE HEROICISATION

Despite a strong sense of intersubjectivity and intimate relating inherent to the companionship, the dog is often exclusively heroicised to the estrangement of its blind partner. Such a heroicisation rests on the contrasting reputations that the dog and the blind person hold in the cultural imaginary: while the former 'is always interpreted as the competent member, as the one with the "good reputation"', the latter is deemed 'as incompetent, as the one with the "stigma"' (Michalko 162). The binary based on competence and incompetence emerges from the ocular-centric worldview that pits the partner with sight as having the agentic command and the one without it as having none. Being subject to misconception, the memoirist evinces unease about his prosthetic appearance in the social space: 'Did they see me as someone who's only effectual because of his dog?' (174). He goes on to recount how a student at 'Guiding Eyes' told him that many sighted people believed it was the guide dogs who pulled their companions along and the blind person simply obeyed (174).

The excerpt articulates the second major crisis in Kuusisto's life: the overvaluation of the canine prosthesis at the expense of oneself. It shows that the pre-existing reputation of the dog as an intelligent being tends to eclipse the unique individuality of its human partner. The effectuality of the blind person being dependent on their dog reiterates the cultural notion that posits blindness as a 'lack' requiring a prosthetic fix (Esmail 19). Ironically, the canine-induced liberation from the closet incites stigmatisation of blindness in its own way—fearing which, the subject passed as sighted in the first place—since he projects his corporeal difference in the open. Besides, 'service animals' are often portrayed as capable of 'heal(ing) wounded people' (Kuusisto 181). Countering this misconception, Kuusisto asserts that disabilities never really disappear and clarifies that having a dog helps you re-engage with the world (181). That is, the real contribution of a companion animal, as far as the mobility of a blind person is concerned, lies in its ability to enlarge his outreach. These discourses of lack and of cure envisage the elimination of blindness through canine prosthesis. They deny the blind person's agency and discount his corporeal experience of being blind while intimately relating to his canine other.

The canine heroicisation can be further understood with reference to Julia Miele Rodas's concept of 'satellite' developed in the context of disability care. She states that the '... satellite figure is defined by the relationship to disability, drawn into the circle of influence, attention, power, or "gravity" that develops around the disabled' (Rodas 106). Being a mobility aid, the guide dog forms part of the purview of attraction along with its blind owner but becomes the centre of attention in the public eye. However, the dog does not consciously thrust itself in the limelight; rather, it is the world that imposes satellite identity upon it. Kuusisto vents his frustration on being invisibilised in the shadow of his canine satellite when he states that it seems as if the canine companion is imbued with intelligence and not the blind person. The dog's dominant position of being in the front translates to heroicisation and the blind person's postposition evokes secondariness.[4] Notably, the satellite identity anthropomorphises the dog and erases its animality. This entails what Rodas calls 'double erasure,' whereby the blind subject is erased by the copresence of his canine companion and the latter is denied of any intrinsic significance other than its prosthetic role to the former (497). It misrepresents the dynamic way in which the blind person and his dog interact while holding their respective agentic capabilities intact.

PAST RESIDUES AND THEIR IMPLICATIONS FOR THE DISCOURSE OF COMING OUT

The discourses of cure, lack and canine heroism force Kuusisto to reassess his dog-inspired self-transformation and re-evaluate its radicalness. In response to these reductive discourses, he seeks to succinctly define what canine accessory does and does not mean. He says that dogs per se do not inject self-transformation, but the freedom of movement that they facilitate enable the subject to filter out his inhibitions and foreground his real self. He also revisits his pre-dog past and distills its relevance to the present. In so doing, he realises that it continues to have a bearing on his canine-inflected orientation to the world. He reflects on how the boy he was will always be a part of the man he is today— a man who has learnt to walk freely only late in life. This indicates that the past does not vanish upon the arrival of the dog but remains latent and forms an indelible part of Kuusisto's subjectivity. Moreover, the exuberance of the boyish self still invokes an active response from the mature adult, evident from his response to the sound of sea lions. He speaks of how all three—man, boy and dog–cross the wharf quickly when they hear the sea lions bark. This shows that Kuusisto remains sensitive towards the dormant instincts of his younger self. He conceives of his boyish self not as a cerebral notion—once real but no more so at present—but rather as a figure with as much tangible presence as the man and the dog. This undercuts his desire to sustain the corporeal memory of his younger self although he has gotten over the fears, anxieties and identity struggles related to his boyhood.

The impulses nurtured during the days of passing define the particular manner in which Kuusisto relates to Corky. The buoyancy that defines this partnership owes its genesis to his outdoorsy disposition honed during his closeted existence. In fact, David See, a representative from the guide-dog school who comes to interview Kuusisto on his application for canine assistance, remarks that he needs 'a dog that can really go places' since he is 'athletic' and has 'geographical ambition' (28–9). This establishes the importance of the blind person's original disposition in determining the selection of his canine companion. In examining the roots of his disposition, Kuusisto realises that the life of passing has, in retrospect, turned out to be a prerequisite to the shaping of the life with the guide dog. He contemplates thus: 'The indomitable will of my former self was still a part of my experience… It took boldness to travel without help. And now… I was a quicker, more refined man of the street' (125).

The passage reads as a reclamation of the 'former self' and an acknowledgement of its residual energies still operating underneath the present existence. Listening and unaided mobility, developed initially as strategies to navigate the world as a sighted subject, mature into a finer aesthetic of dignified, graceful orientation with the companionship of the guide dog. The continuity evident across the author's former and present selves compels us to rethink what it means to come out. Based on Kuusisto's narrativisation, coming out seems to be more a process of re-adjusting one's impulses, attitudes and habits to the openness of one's identity than a wholesale renunciation of the past. As Ellen Samuels observes, coming out is not 'an over-the-rainbow shift that divides one's life before and after the event', but one that demands a continuous negotiation with one's past and present (347). This enables us to fathom why Kuusisto constantly engages with his pre-dog existence to divine and reiterate its positive bearing on his present selfhood.

REDEFINING CARE THROUGH INTERDEPENDENCE

As part of his two-pronged response to the reductive discourses discussed above, Kuusisto affirms that his companionship with his guide dog revolves around a dynamic of trust rather than a dialectic of power and of lopsided dependence. The memoir demonstrates that the blind person's unreserved bestowal of trust functions as a key affective impetus to the efficacy of canine guidance. Hannah Macpherson calls such a bestowal a 'gift of trust' whereby 'visually impaired walkers give up their achieved, independent ability to way-find in order to move through areas of challenging terrain . . .' (141).[5] In other words, the blind person shelves their habitual techniques of self-navigation and gives themselves entirely to the care of their guide dog. For instance, when Kuusisto describes his experience of being expertly led by his dog for the first time, he speaks of not having the time to be worried about any obstacles as the latter moved fast and with confidence (63). The swiftness of the movement drives away his mental worries about the world's unknowns. He believes that his canine companion is 'happy' since he plays his 'part' well and 'relieve(s) her of uncertainty' (68–69). He says that he is responsible for '[k]nowing' the 'destination' and providing '[b]alance' to the movement (69). This shows that the dignity of the blind subject remains uncompromised since the mobility demands his contribution. It underscores the fact that both the man and the dog contribute equally to their coupled navigation: the former with his impetus and the latter with her outreach.

The canine guidance also demands a concession of 'autonomy' from the blind person and requires the admission that 'sometimes the dog will know better than the human' (Couser 195). In other words, while the dog takes its command from its human partner, it will nevertheless resist if the terrain presents a danger for their forward movement. In guide-dog parlance, this trait is called 'intelligent disobedience' (Kuusisto 66). Kuusisto explains how the canine guide will not allow a blind person to cross a street if they feel their companion is not making a safe decision. In such a scenario, the blind person must instantly acknowledge that his canine guide disobeys his dictate for a valid reason and must believe that it is in the interest of their collective safety. This 'blind' faith in one's dog amounts to an 'inter-corporeal generosity' (Andrew). It resists positing the 'guided partner . . . as grateful recipient of assistance' (1166). Kuusisto deploys philosopher George Santayana's concept of 'animal faith', meaning 'instinctive belief, or belief without rational foundation', to refer to his unconditional trust in Corky (156). He admits that the canine guidance made him feel fundamentally confident and open in a way he'd never known (156).

This shows that the companionship opens him up to the canine other and, that in turn invites the latter into an intimate relationality. This inter-corporeal bonding posits agency as 'dispersed' rather than being 'rooted in any one body' (Pemberton 92). The dispersal of agency, however, does not amount to a denial or dilution of agency to the blind person since it forms an integral part of the interdependent dynamic of relating with his canine companion.

The co-dependent nature of the assemblage also manifests in acts of mutual care performed by the companions. The heroic discourse, however, conceptualises care essentially as a canine virtue and fails to take into account the numerous facilitative acts performed by the blind person as part of the interspecies praxis of care. Predicated on the notion that only that which is visible is knowable and credible, it discounts these acts done mostly in the private realm which range from the most routine to the most complex. For instance, the blind person carefully accoutres his dog with guiding equipment such as a harness, belly strap and training collar, and takes care of its personal hygiene (Kuusisto 100–101). He watches out for his dog's bodily safety when navigating intricately difficult terrain—for instance, ensuring that Corky's tail does not get pinched or pulled while passing through certain spaces. At times, letting the dog be itself also functions as a care act. Recognising that Corky might relish different aromas that waft around in the woods, Kuusisto 'let(s) her scent deeply' (173). This shows that the blind person's care extends beyond acts facilitative of effective guiding to encompass gestures empathetic of the dog's unique subjectivity and sensorium.

The companionship also reinforces that fulfilling the comparable needs of one's partner is not a prerequisite to oneself being cared for. As the editors of this volume point out, care in an interspecies relation does not work in the manner of a mere 'exchange' but emerges as an instinctual offering of oneself to the other without anticipating anything in return—an opening born out of a deep affective bond between entwined beings (Aiyadurai, et al. 9). The memoir captures the spontaneous, unconditional caring that the man and the dog lend to each other. First, Kuusisto articulates the selfless nature of his guide dog's caring when he observes that dogs don't expect to be cared for in exchange for helping their companion. That is, canine care is not reactive in a crude utilitarian sense. Similarly, although a blind person's initial motive to look for a dog lies in their need for independent mobility, their relationship quickly turns deeply intimate. One incident from Kuusisto's narrative will illustrate the depth of this intimacy. He attempts to suppress his grief to make Corky's death as light as possible: 'I knew I had to force back my tears because I couldn't let her die to the sounds of my distress' (234). Withholding his tears in this instance is significant since it is intended to let Corky die without regretting that she could not tend to her companion when he is in 'distress' as she always used to. His company at the end, when she can no longer provide care as she is meant to, suggests that he did not view her merely as a prosthesis but as an emotional partner whose passing is beyond bearing.

In sum, the chapter has attempted to read the companionship between a blind person and their guide dog through the framework of interdependence by foregrounding the inter-corporeal relating and the mutuality of trust between the companions. It has proposed a sensitive redefinition of care when the praxis in question involves the blind and the canine—both of whom carry vulnerabilities in their respective ways. In fleshing out the ways in which they form a highly intimate interconnection, it has presented an alternative way of looking at this partnership that is often hierarchised.

Notes

1. I extract this word from Merleau-Ponty's use of the same to imply that the body is an active agent as opposed to an inert matter. He writes: 'It is, then, a certain possession of the world by my body, a certain gearing of my body to the world' (291).
2. Coleman-Brown identifies fear as the 'primary affective component' of a stigmatised identity (155).
3. Tobin Siebers makes an interesting observation in this regard: 'Disabled people who pass for able-bodied are neither cowards, cheats, nor con artists but skillful interpreters of the world from whom we all might learn' (324). Viewed in the light of this statement, Kuusisto's practice of passing can be understood as a 'skillful' interpretation of reality although the need to do so originates in his fear of being found out.
4. On the contrary, the dog depends on its human owner for directions, as Girma, a deaf-blind writer, puts it: 'He takes his lead from me' (103).
5. Macpherson uses this phrase in relation to human guides accompanying their blind companions. Although the context in which it is used is different, I find it useful in defining the relationship between the blind person and their canine guide.

Works Cited

Aiyadurai et al. Introduction. *Ecological Entanglements: Affect, Embodiment and Ethics of Care*, by Ambika Aiyadurai, Arka Chattopadhyay and Nishaant Choksi, Orient BlackSwan, 2022, pp. 1–14.

Andrew, Stevenson. "Dog Team Walking: Inter-corporeal Identities, Blindness and Reciprocal Guiding". *Disability and Society,* vol. 28, no. 8, 2013, pp. 1162–167.

Coleman-Brown, Lerita M. "Stigma: An Enigma Demystified". *The Disability Studies Reader,* 5th ed., edited by Lennard J. Davis, Routledge, 2017, pp. 145–59.

Couser, G. Thomas. "A Personal Post (Post)Script: Further Reading/ Reading Further". *Biography,* vol. 35, no. 1, 2012, pp. 190–97.

Esmail, Jennifer. "'The little dog is only a stage property': The Blind Man's Dog in Victorian Culture". *Victorian Review,* vol. 40, no. 1, 2014, pp. 18–23.

Girma, Haben. "Guide Dogs Don't Lead Blind People. We Wander as One". *Disability Visibility: First-Person Stories from the Twenty-First Century*, edited by Alice Wong, Vintage Books, 2020, pp. 101–03.

Goffman, Erving. "Selections from Stigma". *The Disability Studies Reader,* edited by Lennard J. Davis. Routledge, 2017, pp. 133–44.

Kreilkamp, Ivan. "Anthroprosthesis, or Prosthetic Dogs". *Victorian Review*, vol. 35, no. 2, 2009, pp. 36–41.

Kuusisto, Stephen. *Have Dog, Will Travel.* Simon and Schuster, 2018.

Macpherson, Hannah. "Guiding Visually Impaired Walking Groups: Intercorporeal Experience and Ethical Sensibilities". *Touching Space, Placing Touch,* edited by Mark Paterson and Martin Dodge, Routledge, 2016, pp. 131–50.

Merleau-Ponty, Maurice. *Phenomenology of Perception.* Translated by Collin Smith, Routledge, 1981.

Michalko, Rod. *The Two-In-One: Walking with Smokie, Walking with Blindness.* Temple UP, 1999.

Nishida, Akemi. "Relating through Differences: Disability, Affective Relationality, and the U.S. Public Healthcare Assemblage". *Subjectivity,* vol. 10, no. 1, 2017, pp. 89–103.

Pemberton, Neil. "Co-Creating Guide Dog Partnerships: Dog Training and Interdependence in 1930s America". *Medical Humanities,* vol. 45, no. 1, 2019, pp. 92–101.

Rodas, Julia Miele. "Limited Visibility; or, Confessions of a Satellite". *Pedagogy,* vol. 15, no. 3, October 2015, pp. 493–505.

———. "Misappropriations: Hugh Stuart Boyd and the Blindness of Elizabeth Barrett Browning". *Victorian Review,* vol. 33, no. 2, 2007, pp. 103–18.

Samuels, Ellen. "My Body, My Closet: Invisible Disability and the Limits of Coming Out". *The Disability Studies Reader*, 5th ed., edited by Lennard J. Davis, Routledge, 2017, pp. 343–59.

Siebers, Tobin. "Disability and the Theory of Complex Embodiment: For Identity Politics in a New Register". *The Disability Studies Reader,* 5th ed., edited by Lennard J. Davis, Routledge, 2017, pp. 313–32.

SIX

HUMAN AND ANIMAL

'DESTITUTION' AND 'DIVINISATION' IN THE DISCOURSE AND PRACTICE OF CASTE

Ankit Kawade

Introduction

The invocation or experience of animality is a persistent theme in the discourse of caste, especially in dalit thought and literature. Accounts of animality occur in these writings with mainly two interrelated claims. First, that the caste system fundamentally does not allow those subjected to it an experience of what it is to be a fully human person. Second, it involves the claim that dalit lives are lived in the form of a 'reduction', and the evocative name for such a life is oftentimes the 'life of animals'.

This chapter attempts to study the nature of associations with and of the animal that can be evidenced in the discourse and practice of caste. Association is a key word here for it refers both to association in the realm of language, metaphor or thought, as well as association in the realm of life, existence or experience. Beyond what Andrea Gutierrez calls 'the metaphorical potential of animals' (468) for thought and literature, animals feature in a caste society as human beings' social–natural counterparts such that animals and humans can be said to 'share a common world' (469).

This chapter suggests that the invocations of animality found in the discourse and practice of caste, and especially in dalit thought and literature, constitutes a political–philosophical problem worth thinking about. It is worthy of thought precisely because it positions itself towards assessing the worth of questions such as 'who or what is human?'; 'who or what is the animal?'; 'what are the bounds or limits of a political community?'; 'what is the relation between humans and animals in any given society?'; 'what kind of an ethical treatment is owed to the other, whether that other be a human or an animal?'; 'what is the relation between social and ecological justice?', and so on.

Although a wide range of sources can be consulted to think about the question of the human–animal distinction in the discourse and practice of caste, it is Cyrus Mistry's *Chronicle of a Corpse Bearer* (2012) that this chapter fundamentally relies upon. Mistry's novel is an exploration of the life of an individual from the hereditary 'depressed sub-caste' (Mistry 110) of corpse bearers, known as Khandhias, among the Parsis of Bombay. This chapter focuses on Mistry's and others' usage of the term 'untouchable' to describe the condition of the Khandhias without substantially delving into the question of whether caste or *jati/varna* is a feature of Zoroastrianism or not. It can be situated along with other studies which have attempted to analyse caste-based or caste-like structures among non-Hindu groups in India like the Muslims (Azam), Christians (Thomas), Sikhs (Jodhka), Parsis (Maneck) and others.

The Scope of the Social: Human and Animal in the Caste System

Mukul Sharma writes that 'Caste and nature are intimately and inextricably interwoven in India; and yet their interconnectedness has rarely been a subject of examination' (xvi). The status of the distinction between the human and the animal in the caste system invites us to reconsider the *scope* of this system itself. It is important to reconsider whether the caste system only encompasses human beings, or rather, whether it also encompasses within its operative structure other living creatures which are part of the natural life in the Indian subcontinent. If nonhuman animals can be said to be a part of the scope of the caste system, what are the concrete ways in which the system of caste can be observed to structure the lives and deaths of such animals?

While most studies of caste and untouchability focus upon the relations between human beings situated across the vertical axis of the caste hierarchy, this chapter proposes a methodological and conceptual shift towards considering human animals as well as nonhuman animals as existing within the social imaginary and relationality that constitutes the system of caste in the subcontinent (Govindrajan 11; Kapoor 5). While moving towards such an understanding, it is important to keep the following theses in mind indicating the broad outline of my point of departure:

A. The system of caste encompasses humans as well as animals such that the scope of caste sociality could be observed across the hierarchical order of these two sets of living beings in the subcontinent's 'social ecology' (Guru, "Freedom of Expression" 40).

B. The humans as well as the animals in the caste system are internally divided such that a social hierarchy can be observed to persist within these two sets of living beings. Any analysis which seeks to posit a unified and internally undifferentiated category of humans against a similarly represented category of the animals overlooks the social fact of hierarchy operative across species in the discourse and practice of caste.

C. The question of ecology in India is mediated by the character of social hierarchies or inequalities operative in the system of caste such that it is possible to say that not only the humans, but also the animals shall have to be emancipated from its classificatory regime. The project of the annihilation of caste has an implicative importance for the question of the animal as well as for the question of ecology in India such that it becomes imperative to study 'how caste hierarchies are reproduced by uses of nature' (Sharma xvi).

The following analysis builds upon these theses to explore the scope of caste sociality which encompasses both human as well as nonhuman animals. I seek to analyse here the issue of the extinction of vultures and its concomitant consequences upon the Parsi or Zoroastrian community in Mumbai, by focusing upon the Parsi funeral ritual.

THE SOCIAL CONSEQUENCES OF THE EXTINCTION OF VULTURES

Vultures, along with kites, ravens and crows, are classified as 'scavenger birds'. The funeral ritual among the Parsi community in India is facing a veritable crisis owing to the extinction of vultures. Their extinction has been caused due to the use of the painkilling drug 'Diclofenac' in humans as well as livestock populations (Van Dooren 47; Das, et al. 2). Vultures feeding on such carcasses that have more than trace amounts of Diclofenac suffer a lethal renal or kidney failure. Although this drug was banned in 2006 by the government for use among livestock, the vulture population had already dwindled by around 95% by then and is now at the point of extinction. Asad Rahmani, Director of the Bombay Natural History Society (BNHS), summed it up well when he said, 'Diclofenac is lethal to vultures. It does not matter from where they get it, from a dead Parsi or from a dead cow' (qtd. in Umar). What have been the social consequences of the extinction of the vultures in India, especially for the Parsis?

The method for disposing of the dead among Parsis is known as Dokhmenashini, which is reported to have originated in ancient Persia circa 900CE (Karkaria). The Parsi funeral ceremony involves carrying the corpse to a designated site called 'The Towers of Silence', or Dakhma, where the corpse is laid down on an elevated roof-structure. The process involves exposing the corpse to the rays of the sun and then letting vultures and other scavenger birds feed upon it till only its bones remain. The bones are subsequently pushed into a pit built in the centre of the Towers of Silence.

Zoroastrian theology considers elements like fire, earth, and water to be sacred while stipulating dead matter as defiling, polluting, or as a source of contamination (Williams 158). The corpse cannot be buried, cremated or drowned in water as it is seen as polluting or contaminating the sacred elements of earth, fire and water. The extinction of vultures has meant that the Parsi community has had to install solar concentrators near the Towers of Silence in Mumbai to redirect and intensify the sun's rays for quicker decomposition of corpses (Srivastava). However, such a method hardly proves to be useful after sunset and during the monsoons, in which case the corpses take much longer than usual to decompose (Umar). Moreover, as the corpses take much longer to decompose, the air surrounding the Towers is filled with the stench of decomposing corpses, so much so that an ozone generating machine worth $30,000 had to be installed near the Towers by a US-educated Parsi chemical engineer to combat the stench (Srivastava). The extinction of vultures has led certain members of the Parsi community to opt for electric crematoriums or burials. They fear the impending extinction of over a millennium-old funeral ritual which has been a key marker of the identity of the Parsi community (Umar; Karkaria).

It may be claimed that the social implications of the extinction of the vultures have been the most drastic for the Parsis in India. In much of the commentary that has ensued after these implications came to light, the theme of the ecological superiority of this funeral practice over other forms of funeral ceremonies has been emphasised (Van Dooren 49–53; Karkaria). Nevertheless, a curious blind spot in such commentaries is the role that has been historically played by Parsi corpse bearers, or Khandhias, in the funeral rituals. The extinction of vultures has meant that their working conditions have become far harsher and inhumane than they were before, even as one underlines the fact that corpse bearing was always in the nature of a hereditary occupation among the Parsis in India because of which 'Khandias have become the "untouchables" of an otherwise casteless community' (Bhutia).

It is instructive that Lhendup Bhutia uses the term 'untouchables' in his report to indicate the exclusionary practices of the dominant Parsis against the Khandhias. Bhutia further writes, 'It is rumoured that what Khandias do can be so gruesome that it cannot be undertaken without the aid of alcohol as a calming agent' (Bhutia). Bhutia's report reveals the element of fear and danger in working as a corpse-bearer. One of the corpse-bearers that he interviewed claimed that every other chore of his work pales before the fear of entering the Dakhma. It was not the sight and smell of half-eaten corpses and bones that caused these corpse-bearers so much dread, but the living vultures present in their immediate proximity who were more than eager to pick and feast on human flesh. Bhutia further points out an implication of the extinction of vultures upon the work of the Khandhias:

> With the absence of vultures, crows became more abundant. But unlike vultures, they would picked [*sic*] at small parts like the tongues and eyes. Sometimes they would fly off with body parts like fingers or toes, which would occasionally be found dropped in balconies and so on. Khandias would then have to go to these buildings to retrieve those parts.

The commentary over the fear of disappearance of the Parsi funeral practice of Dokhmenashini underlines themes which should give us pause concerning the ethical–political issues of ecology in India. It raises the following questions: how do we understand social practices which may be ecologically sound and socially unjust at the same time? Does the ecological soundness or efficiency of a social practice imply a less harsh assessment of the way it involves social inequality for those who are tasked with the burden of sustaining it? Dokhmenashini, along with the extinction of vultures, allows for us to face exactly these kinds of questions once we consider its social consequences for the life and death of the corpse-bearing Khandhias.

Cyrus Mistry's *Chronicle of a Corpse Bearer* (2012) is used here to study not only the social consequences of the extinction of vultures in India for the Parsi community, but also to think about universal questions surrounding life and death, equality and justice, faith and ecology. The Parsi funeral method raises the following philosophical questions: Is it possible to die (and to live) without doing an injustice to the environment? Is it possible to die (and to live) without doing an injustice to human beings? What happens when these twin demands of justice towards the environment and towards human beings come into conflict with each other?

EMBODIMENTS OF DESTITUTION

Perhaps the most striking feature of the responses of both the vocal members of the Parsi community as well as journalists, commentators, and scholars who covered the issue of the extinction of vultures in India was the scarce, if not entirely absent, reference to the Khandhias tasked with carrying the dead to the Towers of Silence. Indeed, the crucial subtext of the crisis in the Parsi funeral ritual is not only the extinction of vultures but also the dwindling number or the lack of availability of corpse bearers among the community (Gupta). The economic standing of the Parsi community also means that more and more Khandhias refused to carry forward the profession of being corpse bearers and have transitioned to other professions across several generations. The extinction of vultures made for an inoffensive and innocent representation of the precarious condition of Dokhmenashini when it was already reeling under the crisis of not being able to maintain the hereditary subcaste of 'untouchable' corpse bearers. It is a cruel irony that it took the extinction of vultures and not the centuries' old oppressive living conditions of the corpse bearers for orthodox Parsis to reconsider the operative procedure of their funeral method.

Cyrus Mistry's novel is part social history, part fictional representation of the life of the Khandhias as explored through the narrative voice of its protagonist Phiroze Elchidana. Set in the Bombay of the 1940s, the novel outlines a most intimate portrayal of the practices of the Parsi community with an unsparing attention towards the lives of the Khandhias, raising important questions around the politics of dignity and labour as well as the politics of living and dying.

In two moments occurring in the very first chapter of the novel, Phiroze claims two forms of embodiment—that of the corpse and that of the animal. Such a claim of embodiment acts as a non-literal and affective representation of the actuality of an oppressed existence, drawn through the bodily figuration of the corpse and the animal. This is a highly recurrent phenomenon in the discourse of caste, especially dalit thought and literature, where dalits are attributed with being (or they themselves claim to come to feel like) a corpse and/or an animal.

Phiroze says in the first instance, 'One side of my head was throbbing, raw; felt a bit like a corpse myself' (Mistry 11). He also says that he feels like a 'pack animal'. Between these two, one may note that death and animality are the two most important bodily signifiers used to represent the existential situation of the life of the untouchable in the discourse of caste, where the claim of not feeling like a human is expressed in the 'negative language' (Guru, "The Idea of India" 38) of feeling reduced to an animal and/or a corpse.

The second quotation is indicative of the themes of labour, contempt, and lack of dignity of the work of the corpse bearer. Phiroze's account of animality is mediated not only through the aspect of labour and the harsh 'physical strain' that it involves, but more importantly, the moment of 'feeling-like-an-animal' is caused by social contempt which equates the impurity attached to the dead body in Zoroastrian theology with the body of the corpse bearer, such that it becomes possible for Phiroze to claim that just like the dead body, he too is never touched.

Mistry's description of the work of Khandhias also invites a reconsideration of the ecological romanticism with which the Parsi funeral ritual is portrayed at the cusp of the extinction of the vultures. The sheer difficulty of the labour of the corpse bearer, rather than soliciting any amount of positive recognition from the Parsis either in the form of respect or in the form of higher remunerations and humane working conditions, only invites condescension, abuse and contempt.

Mistry's descriptions provoke us into thinking about the embodiment of destitution that untouchables feel they have acquired or been reduced to within the system of caste. As noted above, two such forms of embodiment occur in the discourse of caste—the corpse and the animal. These can be found not only in Mistry's work but also in other works of dalit thought and literature. For instance, in Shankarrao Kharat's Marathi short story titled 'A corpse in the well', the village Mahar named 'Anna' is ordered, against his will, by a police constable to pull out a corpse from a well. Upon entering the dilapidated well, Anna realises that there are snakes in it. Kharat (5) writes how, '[w]ith the agility of a snake, he [Anna] swiftly climbed the rope', comparing his agile bodily movement in the well with that of the snakes. In Annabhau Sathe's Marathi short story titled 'Gold from the grave', the protagonist Bheema is shown as extracting pieces of gold jewellery from the charred remains of corpses in a Mumbai graveyard and whose only company are jackals as they look for dead flesh in the night. The climax of this story shows Bheema gruesomely having to fight a jackal as they both compete for the remains of the corpse.

Such brutal, and sometimes fatal, encounters with animals in the context of the forced hereditary duty of corpse bearing upon dalits raises the following questions: what is the relation between destitution and death? What is the relation between destitution and animality? What explains the predominance of dalit (oppressed) experiences of what might be termed 'human animality', 'living death' (Baxi), or 'social death' (Patterson)? These questions arise particularly when one focuses on the articulations of subaltern and anti-caste experiences within the discourse of caste. However, other sources in the discourse of caste raise a different problem for us which logically corresponds to and supplements the kinds of experiences produced and sustained by the affect of destitution. It concerns the problem of the affect of divinisation in the caste order and here a detour into classical Hinduism becomes necessary.

EMBODIMENTS OF DIVINISATION

If destitution, death, and animality are the predominant affects that define the subaltern or dalit experience within the system of caste, what about the corresponding experiences of non-subaltern and non-dalit embodied subjects within the same? What affects characterise their experiences of the social order of caste? It is proposed that a peculiar affect of divinisation persists within the caste order that characterise the experiences of the dominant. How this provides us with a different imagination of animality shall also be looked into.

Divinisation is an important affect for the caste order because it carries the mark of the distinction between the sacred and the profane, or the pure and the impure, that separate the existence of a dominant social group from that of others (Douglas). However, the thesis of this section is that the divinisation of, for instance, the brahmin within the classical discourse of Hinduism is also mediated through animality. Here the cow becomes the most important animal to consider. Indeed, the cow and the brahmin are existential correlates within brahmanic theology, whose lives are positioned at equal levels of divinity. The significant part of this correlation is that the murder of either invites the harshest form of punishment—which is the death penalty—since they are both embodiments of divinity.

In this regard, Gutierrez, in their analysis of Dharmashastra literature, refers to the work of Agamben and points out that 'our relations with other animals depend on our relations with the divine' (467). Gutierrez's critical engagement with Agamben is here as impeccable as it is inadequate. By referring to 'the divine' as the point of contact in the relation between humans and animals, it not only misses the fact of internal differentiation or hierarchisation within both the animal and the human worlds as suggested by the discourse of caste, but also risks losing sight of the peculiar experience of destitution that befalls the subaltern within the dharmic ritual/social order. While a consideration of the relation between humans and animals through the connecting thread of the divine (and here the relation of the cow with the brahmin is a highly apposite example) may be plausible, it does have its limits as it does not encompass the experiences of the subaltern within this theoretical framework. One may suggest that in addition to looking at human–animal relations in the dharmic order through the category of the divine, one may as well productively look at it through the category of destitution. For the category of the divine as mediating human–animal relations in the dharmic order constitutes a brahmanic view of the problem, in that it only applies to instances of the simultaneous divinisation of the brahmin along with the cow. However, a subaltern or an anti-caste view of human–animal relations in the caste system would attempt to study these relations through the category of destitution, such that it is possible to understand the embodiments of destitution that are prescribed for a section of the animals (such as the dog) as well as humans (the chandala) in canonical Hinduism. In this regard, the following point Gutierrez makes may be instructive:

> Lower-caste peoples frequently appear in dharmic groups with "lesser" or smaller animals, such as dogs, crows, and insects. The penance for killing animals like crows or dogs was the same as for killing a human Sudra, the lowest of the four classes.... If the dog is often equated with outcastes, Brahmins' ontological grouping with cows is equally prevalent. One must observe the same purity rules for someone who died while protecting a Brahman *or* a cow. (475)

Gutierrez's view helps us understand the system of caste as encompassing humans as well as animals into an 'ontological grouping' that is essentially hierarchical in character. And crucially, it bears noting that a particular class of humans are grouped together with a particular class of animals whereby some humans are made to feel less than human or subhuman or at the receiving end of inhuman treatment, while some humans feel more than just human, that is, superhuman, godly or divine. In the caste system, the affects of divinisation and destitution 'supplement' each other and are mediated by a differential invocation of animality in each case (Sarukkai 195).

DIVINISATION OF THE DESTITUTE, OR, HOW THE DESTITUTE AND THE DIVINE SWITCH PLACES IN THE CASTE SYSTEM

Establishing an interpretative framework where affects of destitution and divinisation (which are mediated through invocations of animality) could be observed as supplementing each other, allows us to study responses towards the reform of social relations within the caste system which have (explicitly and implicitly) made use of such affects. It is proposed that particular kinds of attempts at reforming or rearranging the social relations extant within the caste system, especially relating to the condition of untouchability, effectively seek to divinise the destitute. Put differently, programmes of the 'amelioration' of untouchability often seek to divinise or make sacred and holy the body, labour, and life of the destitute untouchable. One may go a step further and propose that this logic of divinisation of the destitute both exemplifies the scope, as well as underlines the limits of the Hindu response to the problem of untouchability. Mahatma Gandhi's concept of 'harijan' (a term no longer in respectable use, meaning 'God's children' or 'God's people') was the most wide-ranging attempt towards solving the problem of untouchability through divinising the destitute (Guru, "Gandhi's Conception" 12; Kumar 457–61). However, Mistry's novel outlines a highly thought-provoking practice through which such a divinisation of the destitute has been institutionalised in Zoroastrian theology and social practice.

Phiroze, the protagonist from Mistry's novel, is a 'nussesalar'. These are corpse bearers who have to undergo several weeks of training at the Parsi fire temple, after which their initiation ceremony is held by a Parsi priest. Beyond their place in the ritual practices of Parsi funerals, their place in Zoroastrian theology and metaphysics is much more thought-provoking.

The nussesalar in Zoroastrian theology and social practice is an instructive example of the divinisation of the destitute. Amongst Hindu social practices as well, modern attempts at reforming the caste system take the form of making certain individuals from among the untouchables into priests. However, the difference between such Hindu-reformist attempts and the Zoroastrian nussesalar is that he (only male members can officially be corpse bearers) is not a generic priest who officiates at the fire temple. He is an *untouchable* priest as it were, 'a glorified untouchable' as Phiroze calls himself in the novel, only able to officiate at funeral ceremonies, further solidifying the Khandhias' association with death in social practice (Mistry 18). Mistry's narration reveals that the real task of the nussesalar is not merely to protect the purified corpse from the touch of 'overly emotive mourners', but to protect '*the living*' from the embodiment of impure matter i.e., the corpse. This is suggestive of how the appearance of purity among a class de facto requires a substrate of impurity which is supplemented away into another class of human beings—or even corpses—as this particular case demonstrates. Mistry further makes Phiroze say in the novel: 'Through prescribed ablutions, prophylactics and prayers, I'm supposed to protect the general populace— and myself— from the noxious effects of the dead' (17). The nussesalar is supposed to take all the toxicity into his own being.

In this regard, Gopal Guru makes a highly persuasive point about the 'outsourcing' of dirt and impurity onto a class of untouchables so as to continue to maintain the systematic appearance of purity among the higher caste groups (Guru, "Archaeology" 219). The nussesalar is the sacrificial entity par excellence in this regard, whose living body is ritually and ontologically made to 'absorb' the metaphysical impurity accruing from the dead body or the corpse. Zoroastrian theology portrays such a work as a highly noble example of service, where a class of human beings become ontological carriers of systematic and metaphysical dirt and thus save the entire community from its 'evil' effects. Phiroze explains: 'In return ... the scriptures promise, his soul will not be reborn.... What the scriptures forgot to mention, though, is that in this, his final incarnation, his fellow men will treat him as dirt, the very embodiment of shit ... untouchable to the core' (Mistry 17–18).

Concluding Remarks

What are the political–philosophical and social–ecological implications of what we have called the 'divinisation of the destitute'? The first aspect that strikes any observer of such a practice—beyond its ideological and political ingenuity—is that such a divinisation hardly makes a structural dent into the actual conditions of the life (and death) of the destitute. Not only does this technique not solve the problem of untouchability, rather it prolongs it and what is worse, it prolongs untouchability by giving its practitioners a feeling of pride at having a good conscience. As Mistry notes Phiroze remains 'untouchable to the core' in spite of being ordained as a death-priest, as the 'Lord of the Unclean' (Mistry 17–18).

Finally, one may end with the following questions so as to further investigate the implications of this process—what kind of social relationships are envisaged by the puritan technique of divinising the destitute? What kind of associations are possible between the divinised destitute and mortal others? Does divinisation render them as 'ex-destitutes', just as names such as scheduled castes (SCs) and/or harijan are supposed to render untouchables into 'former untouchables' or 'ex-untouchables'? Must one always divinise nature, women, animals, and untouchables in order to 'protect' them? Is divinising the unequal a *just* response to the problem of inequality?

Acknowledgements

I would like to sincerely thank Ambika Aiyadurai, Arka Chattopadhyay, Vivek Narayan and Krishnaunni Hari for their invaluable comments on an earlier draft of this paper.

Works Cited

Azam, Shireen. "Blind Spots: Caste in Contemporary Muslim Autobiographies". *The Caravan*, May 2020, vol. 12, no. 5, pp. 88–95.

Baxi, Upendra. "Restoring 'Title Deeds to Humanity': Lawless Law, Living Death, and the Insurgent Reason of Babasaheb Ambedkar". *Conversations with Ambedkar: 10 Ambedkar Memorial Lectures*, edited by Valerian Rodrigues, Tulika Books, 2019, pp. 134–46.

Bhutia, Lhendup G. 'Khandias: The Keepers of Doongerwadi'. *Open*, 15 Jul. 2015, https://openthemagazine.com/features/living/khandias-the-keepers-of-doongerwadi/. Accessed 26 May 2021.

Das, Devojit, et al. "Diclofenac Is Toxic to the Himalayan Vulture *Gyps Himalayensis*". *Bird Conservation International*, vol. 21, no. 1, 2011, pp. 72–75.

Douglas, Mary. *Purity and Danger: An Analysis of the Concepts of Pollution and Taboo*. Routledge, 1966.

Govindrajan, Radhika. *Animal Intimacies: Interspecies Relatedness in India's Central Himalayas*. U of Chicago P, 2018.

Gupta, Jayanta. "No Pall Bearers, Parsi Man Denied Ritual Farewell". *The Times of India*, 31 Oct. 2017, https://timesofindia.indiatimes.com/city/kolkata/no-pall-bearers-parsi-man-denied-ritual-farewell/articleshow/61350471.cms. Accessed 26 May 2021.

Guru, Gopal. "Archaeology of Untouchability". *The Cracked Mirror: An Indian Debate on Experience and Theory*, by Gopal Guru and Sundar Sarukkai, Oxford UP, 2012, pp. 200–22.

———. "Freedom of Expression and the Life of the Dalit Mind". *Economic and Political Weekly*, vol. 48, no. 10, 2013, pp. 39–45.

———. "Gandhi's Conception of Harijan". *Economic and Political Weekly*, vol. 54, no. 40, 2019, pp. 10–12.

———. "The Idea of India: 'Derivative, Desi and Beyond'". *Economic and Political Weekly*, vol. 46, no. 37, 2011, pp. 36–42.

Gutierrez, Andrea. "Embodiment of *Dharma* in Animals". *Hindu Law: A New History of the Dharmaśāstra*, edited by Patrick Olivelle and Donald R. Davis, Jr., Oxford UP, 2018, pp. 466–79.

Jodhka, Surinder. "Changing Manifestations of Caste in the Sikh Panth". *The Oxford Handbook of Sikh Studies*, edited by Pashaura Singh and Louis A. Fenech, Oxford UP, 2014, pp. 583–92.

Kapoor, Shivani. "'Your Mother, You Bury Her': Caste, Carcass and Politics in Contemporary India". *Pakistan Journal of Historical Studies*, vol. 3, no. 1, 2018, pp. 5–30.

Karkaria, Bachi. "Death in the City: How a Lack of Vultures Threatens Mumbai's Towers of Silence'". *The Guardian*, 26 Jan. 2015, https://www.theguardian.com/cities/2015/jan/26/death-city-lack-vultures-threatens-mumbai-towers-of-silence. Accessed 26 May 2021.

Kharat, Shankarrao. "A Corpse in the Well". *A Corpse in the Well: Translations from Modern Marathi Dalit Autobiographies*, edited by Arjun Dangle, translated by Priya Adarkar, Orient Longman, 1992, pp. 1–6.

Kumar, Aishwary. "The Ellipsis of Touch: Gandhi's Unequals". *Public Culture*, vol. 23, no. 2, 2011, pp. 449–69.

Maneck, Susan Stiles. *The Death of Ahriman: Culture, Identity, and Theological Change among the Parsis of India*. 1994. U of Arizona, PhD dissertation.

Mistry, Cyrus. *Chronicle of a Corpse Bearer*. Aleph Book Company, 2012.

Patterson, Orlando. *Slavery and Social Death*. Harvard UP, 2018.

Sarukkai, Sundar. "Phenomenology of Untouchability". *The Cracked Mirror: An Indian Debate on Experience and Theory*, by Gopal Guru and Sundar Sarukkai, Oxford UP, 2012, pp. 157–99.

Sathe, Anna Bhau. "Gold from the Grave". *Homeless in My Land: Modern Marathi Dalit Short Stories*, edited by Arjun Dangle, translated by H.V. Shintre, Disha Books, 1992, pp. 64–69.

Sharma, Mukul. *Caste and Nature: Dalits and Indian Environmental Politics*. Oxford UP, 2017.

Srivastava, Sanjeev. "Parsis Turn to Solar Power". *BBC*, 18 Jul. 2001, http://news.bbc.co.uk/2/hi/south_asia/1443789.stm. Accessed 26 May 2021.

Thomas, Sonja. *Privileged Minorities: Syrian Christianity, Gender, and Minority Rights in Postcolonial India*. U of Washington P, 2018.

Umar, Baba. "Without Vultures, Fate of Parsi 'Sky Burials' Uncertain". *Al-Jazeera*, 07 Apr. 2015, https://www.aljazeera.com/features/2015/4/7/without-vultures-fate-of-parsi-sky-burials-uncertain. Accessed 26 May 2021.

Van Dooren, Thom. *Flight Ways: Life and Loss at the Edge of Extinction*. Columbia UP, 2014.

Williams, Alan. "Zoroastrianism and the Body". *Religion and the Body*, edited by Sarah Coakley, Cambridge UP, 1997.

AFFECTIVE EXPRESSIONS

SEVEN

EXPRESSIVENESS AND AFFECT

EVERYDAY POETICS IN NATURAL LANDSCAPES

Nathan Badenoch & Toshiki Osada

AN EVOCATIVE ETHNOGRAPHY OF EXPRESSIVE MEANING

The 'affective turn' in the social sciences has opened up space for the treatment of feelings, sentimentality, intimacies and emotion in anthropology. With a framework of affect, we can challenge a range of convenient but simplistic dualisms that do not stand up to empirical investigation. Lutz and White (1986) have commented on the potential of dismantling the juxtapositions of mind and body; self and other; private and public; personal and political; and agency and structure. To the list of dualisms requiring a more evocative ethnography we can add human and nature. Kohn (2013) in his book *How Forests Think* begins the discussion of the possibility of recognising an ecological semiosis not limited by the conceptions that define the human world with a Runa Quichua expressive word depicting the sound of a wounded peccary falling into a river—*tsupu*. He argues that as an expressive, also known as ideophone, defined by the imagery evoked by the sound uttered by the narrator, *tsupu* seems to be somehow outside of language. He argues that the forest is made up of a multitude of selves that interact with humans. It is almost as if the forest has a voice that it can use to narrate these interactions without using human language.

To Kohn, these interactions are semiotic, but not language-like. He asserts convincingly, that *tsupu* depicts the peccary plunging into the water because that is what it sounds like, and that is the imagery that it evokes among speakers of Runa Quichua. For the speaker and the hearer, it feels like *tsupu*. He explains that even people who don't understand the language will feel *tsupu* when it is explained to them. This must be what native speakers feel, but the sound–meaning mapping that makes these words special are not arbitrary in spite of

Saussure's famous claim. However, it is also not universal. Gérard Diffloth commented early on that expressives have their own meaning (Diffloth, "Expressive Phonology") and phonology (Diffloth, "Notes on Expressive Meaning"), suggesting that they are used by speakers in an alternative modality in opposition to the prosaic. He also pointed out that assumed universals of sound symbolism are often disproved when new data is brought to light. The aesthetics of expressive language, or a poetic register, is heavily mediated by linguistic culture. Nuckolls argued convincingly that expressives 'sound like life' (*Sounds like Life*) and it can be argued that expressives also 'feel like life' (Badenoch, et al. 3) for all involved in the linguistic exchange. In this paper, we argue that expressive language is used to achieve affect—an important component of feeling and emotion that is expressed not only as an individual's experience of the 'sensuous surroundings' (Abram, *The Spell of the Sensuous*), but as an emotional implication of expressive performance on hearers.

The study of expressives, ideophones and mimetics has advanced in recent years from the uneasy fringes of linguistics to a more confident intellectual space encouraged and expanded by a range of ethnographically motivated approaches (see Choksi; Badenoch, et al. for recent examples in the South Asian linguistic area; see Badenoch, "Sticky Semantics"; Rudge; Bermudez for recent work in other areas). Dingemanse ("Advances in the Cross-Linguistic Study of Ideophones") defines ideophones—which we choose to call 'expressives' following the tradition of research profoundly influenced by Diffloth in South and Southeast Asia—as marked words that depict vivid sensory perception. Nuckolls sees ideophones as 'expressions that are used to simulate, through performative foregrounding, the salient processes and perceptions of everyday life experience' ("To Be or Not to Be" 131).

Expressives are used by speakers to add poetic depiction to prosaic speech—showing rather than telling, depicting rather than describing. Analysis of expressives and ideophones was long fixated on the peculiar relationships between sound and meaning. It has been asserted recently that they are not simply aesthetic embellishment but that they are central to communication (Dingemanse, "Ideophones and the Aesthetics of Everyday Language"). In the literature on Native American language ideologies, ideophonic usage in poetry shows how poets are driven by a preference for aesthetically motivated language that displays the speaker/writer's thinking, motivating the hearer/reader to see, think and act (Webster). The prominent role of metaphor in emotion language suggests that, in languages like English there is a preference for descriptive language rather than a marked expressive language (Kövecses).

Nonetheless, expressive meaning remains notoriously sticky and difficult to pin down. In Santali (spoken in several states across eastern India) *sə̃ĩ-sə̃ĩ* can depict water beginning to boil, an arrow flying through the air, or a snake striking someone. Consultants' explanations are often long and remarkably vivid, but they are more than pairings of sound and experiential imagery (Badenoch, "Sticky Semantics"). There is a strong element of affect achieved through the use of these words, creating links between speakers that are reinforced by shared norms, aesthetics and imaginations. Iconicity means that, to the speakers, these words resemble what they describe but through expressive affect these words also resemble how speakers feel about them.

Spinoza's 'affectus' is a dialogic capacity for affecting and being affected, making it much more complex than the notion of personal feeling. In fact, it may be a 'prepersonal intensity' that defines a relationship between bodies, environment and others. As prepersonal, affect is a non-conscious experience of intensity, a moment of unformed, unstructured potential (Shouse). The interplay between the noun and verb functions of the word 'affect' itself offers an ambiguity that is conveniently captured by the markedness of expressives (and in many languages expressives are considered to be a separate class of words).

Sasamoto has argued that in the previous fixation on the direct or non-arbitrary relationships between sound and meaning, the important factor of context has been ignored. The relevance of the expressive to the linguistic exchange and each actor's position within the social setting, is indeed crucial to success of expressives as affect provoking linguistic acts. Yet, the performance centered view of expressives adopted here requires more than a consideration of context, which has tended to ignore the poetic agency of verbal art. Moving to a notion of contextualisation, our interest shifts to tracing the negotiations of meaning that emerge between participants in social interactions, arguing here that the interactions are not limited to the immediate social words of the speakers. The 'poetically patterned contextualization cues' (Bauman and Briggs 69) that are foregrounded in expressive performance are often taken directly from the nonhuman and even non-animate spheres that interact in the spaces in which speakers and hearers interact.

The result is not simply a reflection of the diverse range of relationships that exist in the socioecological world. Rather, these interactions are created, re-created and transformed by the use of expressives as a type of ecological semiosis (Nöth). Nuckolls pushes the depth of the communication even further, asserting that for the Runa, expressives create a sentiment of shared animacy between not only humans, but among human and nonhuman as well—'by sonically simulating their perceptions of nonhuman nature, Runa enact the salient links between themselves and the nonhuman life world' (Nuckolls, "To Be or Not to Be" 133). One of the difficulties of elaborating expressive meaning has been the interplay between sensory perception and affect. When the affective implications of this type of communication are recognised, we are faced with the possibility that expressives are not the marked forms; in fact, it is the prosaic forms that are marked in the larger scale of semiosis.

Kita has proposed that the meaning of mimetics, as these words are called in Japanese linguistics, is composed of two semantic elements: an analytical dimension and an affecto–imagistic dimension. In the affecto–imagistic dimension, multiple aspects of an experience are represented: affective, motive and perceptive activation. This is Jakobson's 'expressive function', enabling speakers to insert their attitudes towards what they are speaking about. Thus, in offering a 'rendition of an experience itself' (Kita 14–15), the speaker depicts not only images of external sensory perception but evokes feelings and emotions that arise internally. It is assumed that the hearer can imagine and visualise the sensory perceptions through the analytical dimension of meaning, but they are motivated to feel the completeness of the experience through the affecto–imagistic.

People who are overly vivid in their use of exaggerated expressions, gestures or tones of voice might be described as 'affected' (Skoggard and Waterson). This is exactly what expressives do as a matter of course. Through the writing of evocative ethnography, elaborated by Stoller and further developed by Skoggard and Waterson, we strive to unpack what is really experienced and felt in social interactions through 'integration of abstraction and illustration' (Skoggard and Waterson 111). Evocative ethnography confirms the importance of language in this undertaking. Skoggard and Waterson assert that affect

> provides a space for highlighting, examining, and privileging *feelings*, not just thought or the rational mind, and it emphasizes the part played by "structures of feeling" in social activity and interaction. Thus, the study of affect as both noun and verb suggests multiple dialectics: the individual

> and the collective, *habitus* and identity, emotion and relationship, consciousness and action. (112)

We argue that evocative ethnographic accounts of language use, giving full voice to native speakers' linguistic performances, are key to uncovering the entanglements of expressiveness, affect and ecology. From the earliest days of linguistic anthropology, it is held that '[p]erformer and audience shared an implicit knowledge of language and ways of speaking' in which the relationships that underpin these linguistic exchanges, taken for granted by speakers, should become the deliberate focus of analytical efforts (Hymes 6).

In this chapter we introduce and discuss expressives and expressive language in three languages spoken across South and Southeast Asia. We consider the expressive affect achieved by these words in the context of human–nature interactions. First, expressives of rain are introduced to show how the experience of nature is perceived and internalised by speakers of Mundari, an Austroasiatic language spoken in Jharkhand, in eastern India. The semantics of expressives for rain evoke a range of feelings associated with rain in daily life, foregrounding the impact of rain on health, fieldwork and community well-being. The data for this section was collected in the context of producing the *Dictionary of Mundari Expressives* (Badenoch and Osada). Next, we demonstrate how expressives are used in a narrative performance in the Bit language—another Austroasiatic language spoken near the Laos–China border—where the narrator is telling a story of morality based on human–animal interactions. This section is partial analysis of a story recorded and analysed in the village. Finally, we show how intimacy is encoded in the fauna lexicon of the Sida language—a Tibeto-Burman language spoken in Laos and Vietnam—through expressive morphosyntax. In each, we gain insight into the intersections between individual feelings and collective imaginations. The affective resonance of expressives is used to build feeling worlds that bind individuals, communities and ecologies (Kövesces). Perception is gained in its full sense only when it becomes experience, that is, when it becomes aesthetic—there is no perception without feeling.

BEYOND SENSORY PERCEPTION: RAIN IN THE SENSUOUS LANDSCAPE OF MUNDARI

Expressives are an integral part of the linguistic culture of South Asia. They are found commonly across all language families on the subcontinent. It is striking how many common or similar forms are found in 'genetically unrelated languages', indicating that widespread multilingualism and deep linguistic contact have long been a part of the history of the peoples of the region. Yet each language exhibits diverse characteristics of grammar, meaning and aesthetics in use (Williams; Badenoch and Choksi). The ethnographic material here is the product of elicitation and discussion work with native speaker and research collaborator Ms Madhu Purti, to uncover the complexity of expressive meaning. Such 'folk definition' (Dingemanse, "Folk Definitions") provides critical insights into the speakers' motivations, feelings and responses.

The weather is an important frame through which humans experience their natural environment. The Mundari language of Jharkhand, spoken by more than 1.5 million people in eastern India, has a rich lexicon of expressives depicting rain and how people experience it. Mundari expressives are found in reduplicated form, with identical elements or segmental manipulation between the two (Osada et al.).

The ecological landscapes in which people experience rain is full of expressive language that link human, nonhuman and other elements of the inanimate world. In Mundari, the rumbling of sound of thunder before the rains arrive is *gaɽar-guɽur* off in the distance. The sky lights up momentarily with a single flash, *bijiri-bijiri*, of lightning. Rain falling at some distance still from the observer is *sao-sao*. Frogs may start to croak *teɽ-teɽ* in anticipation of the rain, followed by the *taɽad-taɽad* of large toads heard during the rain. Observation of environmental conditions—perceived as sound, light, distance, directionality and duration in these words—is important for daily life, not only the rhythm of agricultural life in rural areas but also the practicalities of getting daily tasks done safely, as well as the convenience of coming and going in mountainous areas.

Once the rain starts, size and rhythm of the falling raindrops is depicted with expressives such as *tab-teb* (scattered rain with large raindrops, soon ends, dripping in many places) and *tab-tib* (scattered rain with normal sized raindrops, dripping in many places).

Rain dripping off a roof expands on this basic pattern of sound imagery with:

tap-tip: water dripping in singular drops
tob-tob: water dripping in big drops in one place
cotob-cotob: large drops dripping with a relatively constant rhythm

citab-cotob: water drops dripping one-by-one, but not in any regular rhythm

It is interesting to note that the croak of the frogs and toads shares the retroflex /*t*/ with the dripping of rain. This reminds us that these expressives are the most obviously iconic, and while they do trace complex semantic mosaics, they are still the most straightforward of the expressive lexicon for rain.

The themes of wetness and cold are mixed in with many expressives of rain. For heavier rain, we find the following expressives that have the common theme of dripping as a result of being thoroughly wet, in contrast to the dripping expressives introduced above which focus on the sound and rhythm of a drop. These expressives foreground of an indexical sense of experience, evoking the social implications of the natural event.

sata-sata: sudden rain; large and heavy drops falling and then stopping, encountered at the onset of the rainy season, as if splashed with a bucket; includes sound

siti-siti: wet and cold rain falling constantly; air is cooled; no sound association

soto-soto: dripping; rain is still falling, or water coming off the roof; body is completely wet

sutu-sutu: dripping; rain still falling; colder than *siti-siti*, but not as wet as *soto-soto*

suru-sutu: clothes wet from getting caught in the rain, can be wrung out

Lighter rain with a misty feel is accompanied with imagery of the different type of wet, also with concrete indexical implications.

pusu-pusu: misting rain; makes the hair wet but not surrounding environment

pusur-pusur: light misting rain; getting things wet; continuous conditions cause wetness

pudui-pudui: light rain; drops slightly larger than misting rain; slowly getting wet

pudul-pudul: light misting rain; does not soak; ‘rain falls like cotton’, like a puff of dust when threshing rice

All these expressives comment on the direct physical contact with the rain. Other expressives for rain evoke imagery and feelings of how one’s activities are impacted by the rain as well. For example, heavy rain may be depicted as *jaɽa-jaɽa*, a sudden rain that falls suddenly and stops like a bucket splashing. This rain is common in July and people doing business in the market will have to take cover and protect their goods, bringing them back out once the rain has stopped. The expressive depiction is based on sound imagery, but in usage the speaker will be concerned with the implications of the time and other effects of this rain. A *jaɽam-jaɽam* rain is sudden, falling heavily and constantly. The final /-m/ indicates continuation, distinguishing it in terms of duration from the previous example. In this rain, one must stay in the house all day and night. While the rain falls, rivers and paddy fields fill up and people may worry about damage in the fields. While people worry about the fields, the mud house floors of their houses may start to erode with the rainwater.

The expressive *biɽaʔ-biɽaʔ* depicts rain slowing down after a heavy downpour, the drops large but perceptibly dispersed. This expressive is often used when one had thought that the rain had stopped but realised that it had just slowed and is still falling. During the heavy rain, one cannot hear the sound of individual drops. Once the rain has stopped, one is able to hear water dripping *biɽaʔ-biɽaʔ* off the roof of the house or from the leaves on the trees. When the rain is *biɽaʔ-biɽaʔ*, the precipitation has slowed but not actually stopped and one is not yet sure if one can go out or should wait a little longer. This includes a feeling of surprise, as well, as it was assumed that the rain had stopped and one could resume work.

Thus, the unpredictability of the weather is deeply encoded in expressives of rain. The feeling of ambiguity regarding ambient conditions may contribute to a perceived sense of semantic ambiguity in the words used to depict it. For example, *jimbiɽi-jimbiɽi* is light and misting rain that continues. One is not sure if the rain is going to get heavier or not, so it is difficult to decide whether or not to go out into the forest to work.

> putukuiʔ halaŋ bir-te senoʔ=ɲ mone-aka-d-a. daʔ jimbiɽi-jimbiɽi gama-ja-d-a-e.
>
> senoʔ-aɲ cii ka
>
> I wanted to go collecting mushrooms in the forest, but the rain is misting and I don't know if I will be able to go.

In this case, the concern is not of a sudden restart of *biɽaʔ-biɽaʔ* rain, but rather the problem of being constantly wet for an extended period of time during a long period of work or travel. Similarly, work is also disturbed by *jipir-jipir* rain that starts and stops, falling relatively heavily at times.

Lasting longer than *jimbiṛi-jimbiṛi*, in this type of rain men will still go to plough the fields because the rain is not so heavy as to impede work. They will not take umbrellas and will wear only a loincloth because they will get soaked in this rain.

Another rain experience is *juru-kuṇḍu*, which is sudden and heavy and will soak people thoroughly. In this case, people will not usually have time to find shelter. The original meaning of this expressive is 'cold' but it is now limited to a depiction of rain. Unable to avoid the rain, the body gets cold because one has no protection and the person caught in it will wrap their arms around the upper torso. The cold element is critical, because in the typical scenario one will be caught a long way from home, raising the risk of catching a cold as one journeys back in a cold, wet state.

> aaŋga-daaŋga-piṛi-re juru-kuṇḍu-ta-n=eʔ gama-ke-d-a.
>
> A sudden, drenching rain fell in the empty field [making us feel cold and lonely].

In this example, the spatial emptiness of *aaŋga-daaŋga* together with the physical cold of *juru-kuṇḍu* rain creates a sensual depiction of loneliness that spans the natural, physical and psychological. Taking this phenomenon further, we can see how expressive meaning can undergo what seems to be a metaphoric transformation through the embodied sensory perceptions and internal imaginations of darkness.

At the onset of the rainy season, *palad-pilid* ('lightning flashing across the sky') in the sky is a sign that the rains are coming in March. When the sky darkens with heavy clouds, *gul-gul*, one can be overcome with a feeling of apprehension about the future.

> daʔ neʔ-ge-ʔ gama-i-a, rimbil=eʔ gul-gul-a-ka-d-a
> when it is about to rain, the sky is enclosed in heavy dark clouds

This feeling is *alae-balae*, an expressive meaning 'troubled mind, concerned about the future'. The term is etymologically not an expressive, because 'balae' is a prosaic noun meaning 'problem; concern'. However, the phrase has been expressivised, because of its strong psychological connotations, by the addition of the chiming element 'alae'. From an aesthetic point of view, it is not only the mapping of sound meaning that enables expressivity but also the rhythmic contours of repetition and rhyme in the form. Yet the expressivity is less than the majority of expressive forms in which there is no etymological transparency, and we find extension of meaning that includes both physical sensation and emotional imagery in descriptive metaphoric application—from the more direct *rua-te alae-balae* ('delirious from fever') to the more abstract *jii-ko alae-balae* ('heart troubled by the desire to see a lover'). In the latter we hear how the emotions of the speaker are infused into the imagery.

These emotional associations are common in the songs that accompany seasonal rituals. We follow the expressivisation of *alae-balae*, further enhanced by its use with a rain expressive. The anxious feeling of *gul-gul* is linked to the element of darkness, as this expressive is also used for the dark color of water in a deep pool. When the sky becomes *gul-gul*, one has the feeling of being submerged or enclosed in a tight space without light. The implications of the rain to come are personal, resulting in specific feelings. This song demonstrates the linkage between the physical environment, a state of mind, and emotions.

cetan-o laari <u>gul-e-gul-e</u>	The skies are dark and imposing above
latar-o laari <u>alae-balae</u>	the heart is clouded below
catar-o ka-m sabana, mai-na	You don't have an umbrella, young girl
sari-go lum-e-ta-n-a.	and your sari will get wet.

The first two lines of the song are affecto–imagistic depictions that tie together darkness and impending rain with an emotional state of insecurity. The parallelism of these two lines links 'above' and 'below' with external and internal states, playing with the final /e/ sound of the expressives. These expressives of rain are part of a larger ideology of poetic performance embedded in language use by Mundari speakers. The linguistic forms range from iconic to indexical depictions that allow for a dynamic interplay of sound, meaning and feeling.

SUBJECTIFICATION: FOREST AS AFFECTIVE SPACE OF HUMAN–ANIMAL COMMUNICATION IN BIT

Kohn's exploration of the Runa expressive *tsupu* opens a window into human–forest relations. This section moves to the upland areas along the Laos–China border to look at the forest as an affective space for communication between humans and animals. Bit is spoken by approximately 2,800 people, and since it is an Austroasiatic language we can say that Bit speakers are related across time and space to speakers of Mundari. Similar to Mundari, the use of expressives is an important part of their linguistic culture (Badenoch "Sticky Semantics"). These words are used in almost all arenas of language in the village and are a powerful poetic tool in narratives.

The Bit now live on the edge of the forest, and until recent shifts to cultivation of rubber for the Chinese market, their livelihoods depended heavily on their knowledge of the forest. The forest is inhabited by many different life forms, and as hunters, the Bit animist conceptualisation of human–animal relations is complex. The following analysis draws on a story told by Sone, a 50-year-old male villager who has spent much time in the forest, hunting, wandering and studying magic. The story is about a man who takes up a challenge with two close friends to see who can become the richest in three years. The main character in the story decides to make his fortune by hunting, but along the way expresses strong feelings of empathy for the forest animals he understands that he must kill, and more importantly, sell.

We join the story midway, with the man on a hunting trip. He raises his gun and takes aim at a deer, ready to pull the trigger and secure many days of good eating, while at the same time moving close to winning the competition with his friends.

ŋɒɒ həə mɛɛn tɛʔ kwaa peɲ	He went into the forest to hunt
pii tii nɨŋ waʔ	He went in the first year
pos hee ʔɨɨm tɛɛ tɛɛ laʔ peɲ	These deer, they came and he was ready to shoot
klʔɛh ŋɒɒ kɔɔ lɛʔ ruup	But [he] pitied him, because he has a face too
pŋaay lɛʔ roklok	He has eyes, round and bright
lɛʔ sŋkɨɨy lɛɛ bah peɲ	He has life, so he didn't shoot
ʔooo yɨm snaat dee ʔoon səʔ	Oh, [he] took his gun and put it down, like that.

After the deer, several types of birds arrive in the forest area where he waits—*mih ruup ŋɒɒ roklok bah coʔ peɲ* ('he looked at his face full of life and didn't want to shoot him'). He is overcome with the same feeling of common animacy reflected in the eyes. After two years, he returns home to his wife, empty handed. 'You haven't gotten anything in the forest? Don't worry, they are smart. I have collected vegetables and we have plenty to eat.' She encourages him patiently, saying *mnaa coʔ sŋkɨɨy coʔ səəm* (hurt–heart–heart–mind; 'don't feel bad, don't feel discouraged'). Normally the phrase *coʔ səəm*, the main element of this elaborate phrase, is used to indicate hurt feelings. His wife uses this expanded construction, indicating that his disappointment goes beyond feelings to include his sense of morality.

Through the use of *sŋkiiy* ('breath; life force'), the narrative links the hunter's concern for the life of the animals to his experience in the forest. However, the key to the affective language is the expressive word *roklok* ('round and bright'), said of eyes (*pŋaay*). Expressives are generally understood to depict vivid sensory perception—in this case, the roundness of the eyes, the clearly visible whites surrounding the iris, the shine and reflection that gives the eyes their special window onto the soul. This first use evokes feelings of empathy and compassion. In the second use of *roklok,* the imagery of the 'eyes bright and full of life' is used to evoke the human-like qualities of the face (*ruup*). This is a word of Pali–Sanskrit origin, borrowed by way of a Tai language sometime in the past, with the original meaning of 'form; shape'. The common Bit word for face is *rŋmaaŋ*, (literally, 'that which is in front') or in poetic situations *pŋaay mus* ('eyes nose'). However, the choice of *ruup* here brings a more elegant and sophisticated, poetic feel to the narration (Badenoch, "Elaborate Expressions") that is indexical of a deeper empathy. Drawing on a Sanskrit term, the narrator inserts a message of compassion that is familiar to non-Buddhist groups living in close proximity to the Buddhist Lao, but not internal to the Bit animist cosmos. Thus, in linking a Buddhist concept to a metanarrative of expressive depiction, the listener is fully inserted into the conflicted moral landscape of the hunter's trials.

Across the languages of the world, expressives exhibit differing degrees of syntagmatic integration (Dingemanse and Akita). In Bit, expressives function outside of the prosaic grammar. They are unique in that they can occur in almost any position within a sentence, as well as outside the units of prosaic syntax. They also use a morphosyntactic paradigm to mark temporal and spatial elements—something that cannot be done with verbs. The lines, realigned to reflect the poetic wordplay, can be interpreted as

pŋaay lɛʔ <u>*roklok*</u> *lɛʔ sŋkiiy* eyes–have–round and full of life–have–breath

Roklok is positioned between *pŋaay lɛʔ* ['[he] has eyes' (eyes-have)] and *lɛʔ sŋkiiy* ['[he] has life' (have-breath)]. The linkages are ambiguous because the expressive is not grammatically tied to any specific prosaic reference. A prosaic construction would normally employ an elaborate phrase of the shape *lɛʔ pŋaay lɛʔ sŋkiiy* ('he has a face and is alive') in an ABAC construction, where the first and third words are the same. However, Sone inverts the subject–predicate of the first element creating an ABBC structure, with the same word in the second and third position, and then inserts the expressive between the two. At first impression, it would seem that from the center of this construction, *roklok* is depicting both eyes and breath of the deer. A poetic analysis would suggest that he has created a marked parallel structure that is anchored by the expressive, appealing to the non-referential sound–meaning imagery within the rhythmic structure of a prosaic grammatical form. According to the normal sense, *roklok* would not depict anything about *sŋkiiy*, but here the familiarity of the eye imagery is transferred to the invisible breath, evoking the hunter's state of mind as he is tormented by the 'hunter's dilemma' (Århem).

Thus, the affective depiction is activated by the expressive linkage of eyes–face–life. This connection is underpinned by a more general encoding of a human–animal connection in the grammar of the language. The Bit pronoun system differentiates gender in the second and third person. Nonhumans are given gender in the same way. In the above text, the narration uses the third person singular *ŋɒɒ* in reference to the hunter in the first line but then uses the pronoun to refer to the deer, with himself left as an implied actor in the narration. In Bit, if the sex of an animal is not known the feminine pronouns are used as default (Badenoch, "Bit Pronouns"). Yet, Sone chooses the masculine to foreground the connection between the hunter and the deer with the pronoun rather than the normal nominal construction for a buck. In the subsequent text, birds are also referred to with the masculine *ŋɒɒ*, which is somewhat unexpected because birds and insects usually take the feminine *koo*. Here he may be further disturbing the moral order in human–animal relations.

When referring to animals the dual and plural forms of the pronouns cannot be used, so a group of animals will take the singular *ŋɒɒ* or *koo*. This means that when the hunter encounters multiple birds in the forest, they must be called *ŋɒɒ* ('he/him') which requires the hunter to relate to them as the individual 'he'. This grammatically imposed individual connection is key to the story because the hunter is competing with his friends and the other two are enriching themselves by stealing and cheating. Our hunter is hunting but unable to kill the forest animals to make himself rich, as he sees this as a betrayal of common animacy (Badenoch, "Fire in the Forest").

The man is then visited by the animals again. This time they are adorned with shining gold and jewels, but the man still is unable to kill them. The narrator informs us that these animals are sent to him because of his virtue; he still refuses to shoot. After three years of unsuccessful hunting, the virtuous but still poor hunter returns home, encountering a dirty and dishevelled old man. The narrator introduces this man with expressives, evoking images of poverty through the face, clothing and smell.

bat hee ŋɒɒ rɔɔt kndɛŋ nɛɛ	And now, he was about to arrive at the road,
ŋɒɒ mɔɔ-mɔɔ rɔɔt boh	close to arriving at the village,
ʔɒɒy mraʔ həə lɛmɛɛn suʔ huul-huul	the old man, stinking,
rehɟah-rehɟiəh raŋtaŋ-ruŋtuŋ ʔiim	filthy face and tattered clothes, he came.

The poor, dirty old man asks to stay at the hunter's house, who takes pity and agrees even though he himself is a poor man living in a small hut and has no fancy food. The man refuses to eat but requests to sleep with the hunter's wife. Here, the description of the old man's repulsive physical appearance is reiterated and embellished with *suʔ huul-huul* ('giving off a stench that fills the room'). The face and the clothes do not actually appear in the text because the expressives directly provide the context. *Rehɟah-rehɟiəh* is an affecto–imagistic representation of a dirty face, while *raŋtaŋ-ruŋtuŋ* describes the condition of old clothes. In his narrative performance, Sone slows down when he comes to these words, pronouncing them carefully to foreground them prosodically for the full affect of an old, lost man moving. From this performative angle, the narration uses three expressives in a row; the first *huul-huul* is semantically linked to the prosaic verb *suʔ* ('to rot' or 'to give off a bad smell'), but after that *rehɟah-rehɟiəh* and *raŋtaŋ-ruŋtuŋ* require no overt topicalisation.

The couple are filled with feelings of compassion for the old man who has no children or grandchildren to depend on, and they agree that he can sleep with the wife. In the morning, when the old man does not come out of the hunter's bedroom, he gets upset and looks into the room. He finds it filled with shining gold but no sign of the old man. Importantly, no expressives are used in the description of this discovery. Rather, the dilemma of the final test of the hunter's compassion and humanity is foregrounded through the use of the repulsive expressives depicting the old man. The sensory imagery is vivid, but the purpose is to set up the difficult feelings of revulsion that the couple would have to deal with, allowing the wife to sleep with another man who is in such bad physical condition. The depiction of that physical condition is not the purpose of the narrative. It is the affect achieved by evoking the feelings experienced by the couple.

The narrator further enhances this dilemma, when he shifts to referring to the hunter and his wife as *cleʔ* ('grandchild'), using the kinship terminology to further deepen the feeling of compassion and empathy towards the old man, called *ʔɒɒy* ('grandparent') since his introduction into the story. There is a wild–domestic element of this story, which is bridged by the old man. He bridges the moral gap with a test in which the family is rewarded for its compassion. The hunter's feeling of empathy for the deer is a different dilemma altogether from the questionable decision to allow the old man access to his wife's bed. In both cases, the moral tensions are made more palpable for the listener through the use of expressives.

To end, we note that title of this story is ambiguous. At the beginning, Sone started with *psiiŋ bun psiiŋ kun* ('the virtuous person')—an elaborate expression that uses the Lao borrowing *bun kun* ('virtuous') with the Bit word *psiiŋ* ('person'). In the process of settling into the story, he also uses the phrase *psiiŋ sat psiiŋ sii*, incorporating another Lao phrase, *sat sii* ('honest'). One of the elders sitting in the room interjects the phrase *psiiŋ lɒh psiiŋ nɒk*, an entirely Bit phrase meaning 'straight person, good person'. The discussion about the title of the story is a consideration of the various ways that the compassion, patience and generosity of the man and his wife are to be related in the narrative (Badenoch, et al.). Expressives are used to evoke physical imagery and make the experiences of the characters more vivid. More importantly, however, expressives produce an affect of the humanlike nature of animals as being worthy of empathy, depicted through the lifeforce shining through their eyes, and the perceived suffering of the poor, old man in his filthy, stinking state. Thus, subjectification of the nonhuman is achieved through the use of the expressive, creating relative affectivity (Kappelhoff and Lehman).

INTIMACY AND PROXIMITY: THE WILD–DOMESTIC CONTINUUM IN THE FAUNA LEXICON OF SIDA

While the two preceding examples have introduced the affecto–imagistic work of expressives in language practice, this final section looks at how expressiveness of intimacy and proximity is linguistically built into a system of ecological knowledge through special means word formation. In the Sida language spoken in Laos and Vietnam by approximately 2,800 villagers, animal names seem to demonstrate the phenomenon of 'the smaller the animal, the longer the name'. In fact, the body size to name length correlation is more complicated; the further from the realm of household life, the more expressive the names become. The longer names are such because they employ various expressive morphophonological means to describe and depict, creating feelings of intimacy in the process. Names for livestock and animals that are frequently hunted are the least expressive; because they are eaten, they are close to the household. Insects found in the forest have multisyllabic, expressive names, while those insects found on people or animals usually have monosyllabic names. Monosyllabic names are opaque in terms of etymology, holding no meaning beyond the identification of the specific animal. Multisyllabic names are more transparent, with meaning derived from patterns of reduplication, alliteration and assonance. This analysis is based on elicitation and discussion in a Sida village in Laos (located just 4 km from the Bit village above) and draws heavily on "The Ethnopoetics of Sida Animal Names" (see for a more detailed discussion of parallelism in the Sida fauna lexicon).

Most Sida words are mono- or disyllabic. The basic shape of a Sida word is CV(T), where C is a consonant, V is a vowel and T is one of three contrasting tones. Word formation processes in the language build on this basic structure, although there are not many long words in the language. The longest word we have recorded is *thɯ̀-tɕhɯ̰̀-thy-ly-la̰-nǿ-thy-ly*, the name of the common spider known as 'daddy long legs'. This name exemplifies the creative and expressive power of Sida word formation and one of its most productive devices, parallel compounding. The name of this insect can be analysed as *thɯ̀-tɕhɯ̰̀=thy-ly≡la̰-nǿ=thy-ly*. The number of bars between the words indicates the hierarchy of nested compounding. The relationship can be visualised as

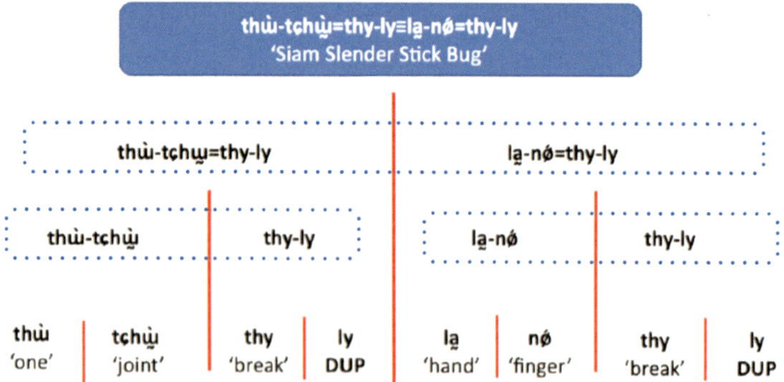

Figure 7.1: Breakdown of the compounded Sida name for a 'daddy long legs'

In this complex compounded name, parallel structures are embedded in three layers. Parallelism of this sort utilises rhyming, alliteration and assonance across syllables to create a sonic quality that appeals to speakers' sense of play. For example, *thy-ly* is a reduplicated form using /l/—an expressive morpheme that plays an intensification role in the prosaic morphology of stative verbs—and the shared /y/ vowel. This construction occurs, in full reduplication, at the right boundary of the main paired constructions. The expressive /l/ also links across the main pair in the *thy-[ly≡la̰]-nǿ* collocation, connecting the expressive /l/ with the lexical /l/ of *la̰-ǿ* ('finger'). The entire name is a descriptive–depictive expression and can be glossed as a short poetic construction: 'one joint snaps, the fingers break'. The reduplicated /l/ intensification also functions as an aspect marker, giving 'breaks' the sense of 'in a broken state'. The complex entanglements of sound and meaning, similarity and difference, across layers of compounding produce a particular affect with the listener.

Parallelism is an important poetic device used all over the Southeast Asian linguistic area (Koret). Its use creates tension between similarity and contrast in paired constructions, as shown in the hierarchy above. In these constructions, mimesis and alterity are not contradictory but complementary (Taussig). Simple reduplication allows speakers to create four-element, parallel constructions (ABBC) that are balanced by the shared element in second and third position. For example, one general term for katydids is *nø-pjú*, and one type of grasshopper is *nø-pjý-pjý-khɐ̀*. After the expressive reduplication at the core of the compound, the lexical word *khɐ̀* ('hard') completes and depicts the physical qualities of its body. Other examples are given below.

phà-jø̀-jø̀-phɯ́	large grey civet	*phɯ́* meaning 'grey'
phà-jø̀-jø̀-nɛ	fragrant civet	*nɛ* meaning 'fragrant'
pjà-tú-tú-pjɐ̰̀	lesser banded hornet	*pjɐ̰́* meaning 'stripe'
pjà-tú-tú-ma̰	Asian giant hornet	*ma̰* meaning 'large'
pjà-ka-ka-nɤ	north-Siam tiger wasp	*nɤ* meaning 'red'

The reduplication sets up a distinctive prosodic feel, anchored by the identical reduplicated sequence at its core. As suggested above, the reduplicated element is not necessarily the most important semantically, and we can thus consider this structure to be an aesthetically motivated reduplication. However, we argue that there is more to this than just creating a balanced, nice-sounding name. Another motivation for this structure may be a mimetic construction of the interrelatedness between the speaker and the insect.

There is evidence in support of this interpretation in the kinship terms used in Sida. In Sida society, maternal relatives require terms of greater intimacy, described as 'names that show more respect'. For example, terms for grandparents encode this intimacy through the same type of ABBC reduplication.

Table 7.1: Intimacy in Sida kinship terms

Relation	*General and Paternal*	*Maternal*
Grandmother	ɘ́-phì	ɘ́-phì-phì-ma̰
Grandfather	ɘ́-pɘ̀	ɘ́-pɘ̀-pɘ̀-ma̰

The maternal forms end with *ma̰*, a morpheme that means 'large' here, but may be etymologically related to the word for 'mother'. The desired feeling of 'respect' is achieved by an interplay of rhythm, sound and meaning. The terms for certain uncles and aunts are distinguished in the same way. What is a semantic distinction of paternal versus maternal kinship seems to be related to the poetic motivations for insect names. We can hypothesise then, that there is an element of intimacy in the parallel reduplication in which human relationships are not entirely different from relationships with nonhumans.

The interplay between the expressive /l/ and rhyming, as introduced in the term for 'stick insect' above, creates different patterns of poetic reduplication evident in many examples. The morphological complexity and aesthetic of /l/ bring enhanced poetic affect to the fauna names, locating us in the forest far from the household. The simple fact of lengthening the names is part of an intimacy marking that is indexed with the increase in syllables and poetics. The following words show the expressive /l/ sound in the third syllable.

pi-sɘ̰̀-lɘ́-sɘ̰̀	insect species with a painful bite	ɘ̰ meaning 'stinging'
pø-khṵ̀-lú-khṵ̀	larvae of longhorn beetle	*khṵ̀* meaning 'curved, curled'
há-thà-là-ma̰	earwig (Dermaptera)	*há* meaning 'iron', *thà* meaning 'pinch'
jó-pá-lá-ma̰	lizard	*jó* meaning 'lizard', *pá* meaning 'to fly'

If we backtrack through to the domestic sphere, this notion of intimacy in the wild is confirmed by two other common species that are part of the household. The monomorphemic insects *mɤ* ('tick') and *hé* ('louse') are both parasites. Similarly, the domestic *ɐ̰* ('chicken') versus the wild *tɕù-í-tɕù-xɯ́-lɯ́* ('scarlet minivet') show how distance along the house–forest continuum is encoded in bird names.

Larger animals are named most commonly with monosyllabic words modified by simple descriptive words, such as *mjɔ̰-ɲý* ('macaque-green') and *ú-ma* ('civet-large'). Many of these are hunter's prey, and thus subject to the type of moral ambiguity discussed in the previous section on the Bit story. In Sida, it is clear that it is simply a matter of the 'smaller the insect the longer the name'. We can summarise the affecto–imagistic semantics of the Sida fauna lexicon as below (Badenoch, "The Ethnopoetics of Sida Animal Names")

Table 7.2: Semantics of the Sida fauna lexicon

Structure	Monomorphemic	Dimorphemic	Polymorphemic
Motivation	Etymological	Descriptive	Depictive
Context	Transparent		Opaque
Position	Practical		Playful
Ecosystem	Domestic		Wild
Intimacy	Distant		Proximate
Mode	Prosaic		Poetic
	⟵	⟶	

The grammar of intimacy brings together human–human and human–nature relationships, providing much more proximate evidence of how expressivity achieves affect in Sida. While the Sida language does not have a large class of expressive words of the type found in Mundari and Bit, speakers create affect through marked sound–meaning mappings in mixed expressive–prosaic constructions. Expressive affect entanglements are encoded in the word formation processers associated with specialised nomenclatures. These embedded indexes of intimacy highlight how ecology and kinship are interlinked in certain lexical areas of social importance to speakers. Expressiveness is cultural context, where relationships are created and reconfirmed in the course of daily language use.

Conclusion

In their discussion of affective communities, Kappelhoff and Lehmann call our attention to Aristotle's thinking about dramatic tragedy:

> Tragedy, then, is an imitation (*mímesis*) of an action (*práxis*) of high importance, complete and of some amplitude; in language (*lógos*) enhanced by distinct and varying beauties; acted not narrated; by means of pity (*éleos*) and fear (*phóbos*) effectuating its purgation (*kátharsis*) of these emotions. (qtd. in Kappelhoff and Lehmann 210)

Expressives are dramatic and emotive performances that tie the speaker and hearer together in the experience of life in their linguistic ecology. Because the sound–meaning mappings are not arbitrary, they work on semiotic principles of iconicity and indexicality; thus, expressives allow speakers to skip momentarily—or at least evocatively disturb the prosaic constraints of—the *lógos* step of this affective process. There is a grammar of intimacy (Webster) that binds both speaker and hearer in the illusion of participation through expressives. We argue that this intimacy is extended into the natural environment that provides the physical, emotional, spiritual and biological context for daily life. At the same time, we hear the 'voices' of the forest and other ecologies speaking to social actors through mimetic representations of being, acting and feeling. Shouse has argued that affect is abstract to the degree that it cannot be realised fully in language. It is reasoned that the body has a grammar of its own that eludes language because of language's inability to 'absorb pulses or discrete stimulations' (Massumi, qtd. in Shouse). We have argued that expressive meaning and the disconnect from prosaic syntax do allow language to make connections between body, environment and community in a way that transcends the multiple dualities of observation–feeling, speaker–listener and human–nature. We see that expressivity and affect are particularly salient for understanding how intimacy is encoded in human interactions with animals in their semiotic landscapes.

We end with a reflection on discussions between a group of researchers that worked for three years on expressives of the South Asian linguistic area. The group was composed of linguists, anthropologists and specialists in the analysis and translation of literature. At many points, as the discussion of meaning crossed between the diverse languages of the region on the back of shared expressive forms, the group members expressed discomfort with the notion of expressive meaning being about the description and depiction of emotion and feeling. The notion of emotion and feeling were too monolithically focused on the subjectiveness of the speaker. What was missing was an entanglement of reflexivity and transitivity in the depiction of these elements of experience; expressives work not only to foreground a special sense of intensity and qualification in the experience, but more importantly, the expressivity evokes a response that entangles the aesthetic, sensual and moral. In this chapter we have examined three perspectives on expressive–affective language use—integration of a set of bodily sensations with emotional reflections in the experience of rain; depiction of shared animacy in interactions with wildlife; and the encoding of cognitive frameworks of poetic intimacy across the domestic–wild spaces of life. These have shown how expressive language links humans and the broader world of animacy, as well as sensuously locating them in an affective ecology of experience.

Works Cited

Abram, David. *The Spell of the Sensuous: Perception and Language in a More-than-Human World*. Pantheon Books, 1996.

Århem, Kaj. "Animism and the Hunter's Dilemma: Hunting, Sacrifice and Asymmetric Exchange Among the Katu of Vietnam". *Animism in Southeast Asia*, edited by Århem, et al., Routledge, 2015, pp. 91–113.

Badenoch, Nathan. "Bit Pronouns in a Northern Mon-Khmer Perspective". *Research in Asian Languages*, vol. 10, 2017, pp. 5–23.

———, and Toshiki Osada. *Dictionary of Mundari Expressives*. Research Institute for Languages and Cultures of Asia and Africa, Tokyo University of Foreign Studies, 2019.

———. "Elaborate Expressions in the Bit Language: Parallelism and Performance in a Multilingual Context". *Journal of Lao Languages*, vol. 1, 2019, pp. 75–87.

———. "The Ethnopoetics of Sida Animal Names". *Topics in Middle Mekong Languages and Linguistics*, Kobe City University of Foreign Studies, 2019, pp. 39–73.

———. "Fire in the Forest: Morality and Human-Animal Relations in a Phong Laan Myth from Northeastern Laos". *Journal of Lao Languages*, vol. 2, 2020, pp. 29–41.

———. "Sticky Semantics: Expressive Meaning in Mundari". *Expressives in the South Asian Linguistic Area*, edited by Nathan Badenoch and Nishaant Choksi. Brill, 2021, pp. 17–34.

———, et al. "Expressives as Moral Propositions in Mundari". *Indian Linguistics*, vol. 80, nos. 1–2, 2019, pp. 1–17.

———, and Nishaant Choksi, editors. *Expressives in the South Asian Linguistic Area*. Brill, 2020.

Bauman, Richard and Charles L. Briggs. "Poetics and Performance as Critical Perspectives on Language and Social Life". *Annual Review of Anthropology*, vol. 19, 1990, pp. 59–88.

Bermudez, Natalia. "Ideophone Humor: The Enregisterment of a Stereotype and Its Inversion", *Journal of Linguistic Anthropology* vol. 30, no. 2, 2020, pp. 258–272.

Choksi, Nishaant. "Expressives and the Multi-Modal Depiction of Social Types". *Language in Society*, vol. 8, no. 44, 2020, pp. 440–47.

Diffloth, Gérard. "Expressive Phonology and Prosaic Phonology in Mon-Khmer". *Studies in Thai and Mon-Khmer Phonetics and Phonology in Honour of Eugénie J.A. Henderson*, edited by Theraphan L. Thongkum, Chulalongkorn UP, 1979, pp. 49–55.

———. "Notes on Expressive Meaning". *Papers from the 8th Regional Meeting*, edited by Paul M. Peranteau, G.C. Phares, and J.N. Levi, Chicago Linguistic Society, 1972, pp. 440–47.

Dingemanse, Mark. "Advances in the Cross-Linguistic Study of Ideophones". *Language and Linguistics Compass*, vol. 6, no. 10, 2012, pp. 654–72.

———. "Folk Definitions in Linguistic Fieldwork". *Language Documentation and Endangerment in Africa*, edited by J. Essegbey, et al., John Benjamins Publishing, 2015, pp. 215–38.

———. "Ideophones and the Aesthetics of Everyday Language in a West-African Society". *The Senses and Society*, vol 6., no. 1, 2011, pp. 77–85.

———, and Kimi Akita. "An Inverse Relation between Expressiveness and Grammatical Integration: On the Morphosyntactic Typology of Ideophones, with Special Reference to Japanese". *Journal of Linguistics*, vol. 53, no. 1, 2016, pp. 1–32.

Hymes, Dell. *"In vain I tried to tell you": Essays in Native American Ethnopoetics*, U of Nebraska P, 1981.

Kappelhoff, Hermann, and Hauke Lehman. "Poetics of Affect". *Affective Societies: Key Concepts*, edited by Jan Slaby, and Christian von Scheve, Routledge, 2019, pp. 210–19.

Kita, Sotaro. "Two-Dimensional Semantic Analysis of Japanese Mimetics". *Linguistics*, vol. 35, 1997, 379–415.

Kohn, Eduardo. *How Forests Think: Towards an Anthropology beyond the Human*. U of California P, 2013.

Koret, Peter. "Whispered So Softly It Resounds Through the Forest, Spoken So Loudly It Can Hardly Be Heard: The Art of Parallelism in Traditional Lao Literature". *Thai Literary Traditions*, edited by Manas Chitkasem, Chulalongkorn UP, 1995, pp. 265–98.

Kövesces, Zoltán. *Metaphor and Emotion: Language, Culture and Body in Human Feeling*. Cambridge UP, 2000.

Lutz, Catherine, and Geoffrey White. "The Anthropology of Emotions". *Annual Review of Anthropology*, vol. 15, 1986, pp. 405–36.

Nöth, Winfried. "Ecosemiotics". *Sign Systems Studies*, vol. 26, 1998, pp. 332–43.

Nuckolls, Janice. "The Sound-Symbolic Expression of Animacy in Amazonian Ecuador". *Diversity*. vol. 2, 2010, pp. 353–69.

———. *Sounds Like Life: Sound-Symbolic Grammar, Performance, and Cognition in Pastaza Quechua*. Oxford UP, 1996.

———. "To Be or Not to Be Ideophonically Impoverished". *Texas Linguistic Forum*, vol. 47, 2004, pp. 131–42.

Osada, Toshiki, et al. "Expanding the Model of Reduplication in Mundari Expressives". *Expressives in the South Asian Linguistic Area*, edited by Nathan Badenoch and Nishaant Choksi, Brill, 2020, pp. 78–99.

Rudge, Alice. "Hidden Likeness: Avoidance and Icononicity in Batek". *Journal of Linguistic Anthropology*, vol. 3, no. 1, 2021, pp. 4–24.

Sasamoto, Ryoko. *Onomatopoeia and Relevance: Communication of Impressions via Sound*. Palgrave Macmillan, 2019.

Skoggard, Ian and Alisse Watterson, "Introduction: Toward an Anthropology of Affect and Evocative Ethnography". *Anthropology of Consciousness*, vol. 26, no. 2, 2015, pp. 109–20.

Shouse, Eric. "Feeling, Emotion, Affect". *M/C Journal*, vol. 8, no. 6, 2005, https://doi.org/10.5204/mcj.2443. Accessed on 16 Feb 2022.

Stoller, Paul. "Ethnography/Memoir/Imagination/Story". *Anthropology and Humanism*, vol. 32, no. 2, 2007, pp. 178–91.

Taussig, Michael. *Mimesis and Alterity: A Particular History of the Senses*. Routledge, 1993.

Webster, Anthony K. *Intimate Grammars: An Ethnography of Navajo Poetry*. U of Arizona P, 2017.

Williams, Jeffrey, editor. *Expressive Morphology in the Languages of South Asia*. Routledge, 2020.

EIGHT

SONGS OF SCRIPT

EMBODIED AFFECT AND INDIGENOUS LANGUAGE LITERACY IN EASTERN INDIA

Nishaant Choksi

In 2009, I first started conducting research on literacy and the creation of new scripts among Santali speakers in eastern India. Santali is an adivasi (Indigenous) language, and part of the Munda branch of the Austroasiatic language family. It is spoken by over 6 million people in the eastern Indian states of West Bengal, Assam, Orissa, Bihar, and Jharkhand, as well as in parts of Nepal and Bangladesh. Most Santals are schooled in dominant Indian languages like Hindi or Bengali, but many have a strong interest in writing and producing literature in their own language and in their recently created script Ol-Chiki.

In order to investigate why Santals enthusiastically embraced writing and publishing in their language, despite the lack of formal education, I initially approached those involved in the business of Santali language publishing. In one of my first interviews, I asked a publisher why writing and publishing had gained so much currency among the Santals and what was its importance in efforts of revitalisation and Indigenous language politics? The publisher admitted that writing was important, but he said, 'We Santals do not learn from books'. Puzzled, I asked if he meant the oral tradition was more important, but he brushed this question off saying that, really Santals learn from 'song and dance'. 'Bengali people learn from an institute, they learn tradition, but for us we don't have a use for that. We bring our tradition forth from the heart; *khuli dhuli khilot' khilot'* ['kicking up dust'; 'kids first learning to dance'].

I thought it ironic at the time that someone devoted to creating a future community of Santali language readers and writers would downplay the importance of books. However, after doing fieldwork in the rural areas where most Santals reside, the publisher's words started to become clearer. I came to learn that more important than the acquisition of language or literacy itself was a feeling called 'raska', which, as the publisher said, comes forth from the heart to express the intensity of communal enjoyment. Talk of these feelings accompanied many activities such as travelling, meeting guests, eating a good meal, enjoying alcohol together with friends, or romantic relations between lovers. However, *raska* was most often talked about with respect to oral performance practices of song and dance (*enec'-seren'*).

Drawing on fieldwork I conducted in the Santal-dominated area of south-western West Bengal between 2009 and 2011, I argue in this chapter that Santali understandings of writing and script exceed both the representation of spoken language and the production and deciphering of graphic, visual marks. I suggest that writing becomes affectively charged through its intimate connection with the sonic and kinetic features of oral performance, such as song and dance. These affective associations with writing have contributed to the popularity of the recently invented Ol-Chiki script, even in a context where proficiency in the script remains low.

Language Maintenance, Writing and the Role of Affect

Recently, anthropologists working with Indigenous communities in North America have emphasized the role of 'feeling' as a way of explaining how continuity is maintained even in situations of widespread language loss. These continuities are cultivated through the pairing of linguistic elements with affectively charged performance practices. David Samuels, working on the San Carlos Apache reservation in Arizona, uses the term 'iconicity of feeling' to describe how community members affectively connect contemporary musical styles sung in both Apache and English with the history and experiences of reservation life. Anthony Webster describes how Navajo poets employ *hodit'sa* (sound symbolism) to evoke similar feelings in listeners regardless of whether the poem is in English, Navajo, or a mixed variety. Barbra Meek and Gerald Carr discuss the role of performance in language revitalisation efforts among the Kaska in northwest Canada, arguing that restored texts in the Indigenous Kaska language only become salient for the community when paired with contemporary and popular performance practices.

These studies' emphasis on performance challenges the idea of loss or revitalisation as anchored to knowledge of a specific linguistic code. Instead, they highlight the importance of the aesthetic and affective component of language, or as Webster writes, an 'intimate grammar', which evokes feelings in the listener that recall communal histories and lifeways. Yet the focus of these studies and others that examine the relationship between affect and text (Besnier; Faudree) remains on the voicing of texts and their reception (reading). In this chapter, I would like to shift the discussion to writing, to focus on how performance interacts with the material production of texts. I will show that for many Santals, oral performance indexes practices of writing (called *ol* in Santali) and endows artifacts like script with affective intensities, where they assume meanings for community members beyond the representation of speech or referential content.

WRITING AND THE TRANSFORMATION OF ORAL PERFORMANCE

Before I began my research, I was aware of colonial-era stereotypes that cast Santals and other Indigenous peoples in India as primitive tribes with a romantic propensity for song and dance (Banerjee). In contemporary India, this stereotype persists in the form of saying groups like the Santals have 'oral'-based cultures, or in a more insulting way, they are 'illiterate'. So, I believed my research on writing and script among the Santals would complicate and challenge these stereotypes. However, my original research questions—which I have written about extensively—were based on a notion of writing as tied to the production of visible text (Choksi, *Graphic Politics in Eastern India*). Yet when I started my fieldwork in the rural areas of eastern India, I saw that what Santals themselves called 'writing' (*ol*) did not fit easily within this narrow conception. Indeed, the practice and valorisation of writing was cultivated in the very space that colonial administrators and many contemporary scholars have characterised as resistant to literacy: the performance space of the village known as the *akhada*.

Akhadas are spaces created at edges of village hamlets or at public events like festivals or weddings specifically for song and dance celebrations (*enec'-seren'*). Villagers created *akhadas* throughout the year, such as during the period of the spring and winter harvest festivals, to celebrate events like India's World Cup cricket victory, or sometimes for no particular reason at all. Both men and women participated jointly in *akhada*s, and they were sites associated with intense communal enjoyment (*raska*). It was said that sometimes, especially in the olden days, *akhadas* were suffused with so much *raska* that sacred spirits would descend and interlock hands with human dancers, disappearing without a trace at the end of the night. Because of the importance of these kinds of celebrations to village life, my interlocutors frequently asked me to record these performances. Even though I thought this was tangential to the focus of my research, I nevertheless obliged and recorded several of what I thought to be traditional performances.

Here is an example of what I understood to be a traditional *lagde* song which, unlike more regimented melodies, are songs that have a looser structure and can be sung at any time of the year. This *lagde* was sung and recorded in a village in Purulia district, West Bengal, just after people returned from the spring harvest festival celebration known as *Baha*.

Dak' chetan poyrani
hod ko metam baha rani,
bahayenam danang rey
okoy kotaay janga dhuri
am samang rey
bahayenam danang rey
okoy kotaay janga dhuri
aam samang rey

The lotus floats on the pond
People call you queen of flowers
Your flower is hidden in the back
who will approach you (dust their feet)
from the front?[1]

This song is ostensibly a poem about a 'lotus' but is also interpreted as a metaphor for a woman's fading youth. When my research assistant (who is a resident of that village) and I were transcribing the song, his uncle (also from the same village and a Santali language magazine editor) who heard us doing the work remarked to us that this song was not a 'traditional' *lagde* but in fact was 'writing' (*ol*). I asked how it could be writing if there were no written texts produced, and he told me it was because of the way it was poetically structured—in particular, its use of eight syllable meter, which is typical of learned poetry in Indo-European languages such as Hindi or Bengali—and associated ideologically with written texts. In fact, he said in that in this region of eastern India, the most common songs of that particular genre were all based on 'writing'. Thus, the production of metrical forms within the *akhada* in fact constituted a *form* of writing that though embodied and sung, distinguished itself from the non-written song tradition which was diagrammed through parallelism and not meter.

For instance, take this song cited from a 'traditional' collection of *lagde* songs:

Opor kuli akhḍa
Namo kuli akhḍa
Akhḍa boro re jomok'
Gitic' jāpic' anjom-aiṅ
Akhḍa boro re jomok'

The upper-lane akhada
The lower-lane akhada
The akhada is in full-swing
Sleeping, I hear it
The akhada is in full-swing

(Hembram 143)

Unlike the song 'Dak' chetan poyrani', which is sung with traditional instruments and expressed through dance, this song does not rely on meter or rhyme to unify its expression. Instead, it depends on slight variation in form within the distich, such as the opposition between the 'upper' (*opor*) and the 'lower' (*namo*), or the insertion of an odd line between two echo lines. In addition, except for the italicised line, which is in Santali, the other lines are in the local Indo-European language (Jharkhandi Bangla variety) showing that these songs in their more traditional, non-written form were also multilingual in nature.

As I found out in my subsequent interviews, many current generation Santali language writers I spoke to became interested in literary careers through their exposure to this form of 'writing' in *akhadas.* For instance, one writer said that when he first started going to school he noticed a clear difference between the Bengali poetry he had to recite and copy in school—which he associated with *akshar* ('written word')—and the sung poetry he routinely heard in his village *akhada*—which was based on *aakar* ('form'). He described *akshar* poetry as that which was based on meter, whose syllables were evenly divided, had end rhyme and was associated with the Bengali language poetry one learned in school. He recited a popular poem by Rabindranath Tagore which all schoolchildren in West Bengal learn:

Kumorpaḍar gorur gaḍi
Bojhai kora kolshi haḍi
Gaḍi chalay Bongshibodon
Shonge je jai bhagne Modon

The bullock cart goes from the potters' neighbourhood
Loaded with earthen pitchers and pots.
Bongshibodon drives the cart,
His nephew Modon accompanies him.
(Tagore 9)

This poetry was associated not with the *akhada* but with the school space, and again was based on meter and not form (*aakaar*) like in the song 'Opor kuli akhda'; nor was it accompanied by the kinetic and musical accompaniment that created affective *raska*. The metrical style and use of end rhyme was part of the socialisation of learning the *diku* (Bengali) language of the caste Hindus and not associated with becoming a Santal (*hod*).

However, according to the Santali author I spoke to, this changed during the 1970s when the region was in the throes of a political movement that sought political autonomy for the Adivasi majority areas in eastern India. During this period, the author heard many songs, both politically themed and otherwise, that sounded to him both 'Santali and not-Santali'. One example he cites is the famous song, often called the anthem of the Jharkhand movement written by Sadhu Ramchand Murmu, which was performed frequently during his childhood in the village *akhadas* and fairs.

Marang buru joy menabon
Mit' horte tadamabon
Bair helec' kate abon
Duk do mabon ṅec' ṅir
Debon tengon adibasi bir do—
Debon tengon adibasi bir

Let us (inclusive) say victory to Marang Buru [the Great Hill spirit]
Let us (inclusive) walk on one path
Let us (inclusive) drive away together all of our worries
Let us (inclusive) stand up, we (inclusive) the Adivasi heroes

In this example one sees that the stanzas are comprised of six lines with a regular eight syllable metrical pattern, which can also be found in the Bengali poetry the Santali writer I spoke to learned to cite in school. It also makes use of end rhyme rather than the parallelism that the writer associated with the Santali form of oral poetry. However, when it was sung, it was sung in a traditional manner—in the space of the *akhada*, accompanied by associated Santal rhythms played out on traditional drums like the *tamak* and *regda* and by Santal-specific dance forms (*enec'*). Thus, even though these songs were not poetically mapped to previous Santal songs, they generated a Santal-specific affect through their enactment in the space of the *akhada*.

This meant that for the Santali writer, they were Santali in that they followed Santali-specific melodies, rhythms, and movements, but not Santali in that they were based on meter and not form. He said that after hearing this he began adapting Santali to meter along the lines of what he learned in school, later turning to other literary genres such as drama and short stories. As anthropologist Anna Stirr has argued for the way the understanding of *jhyare* poetry was transformed in Nepal, in this case 'meter' has become (in semiotic terms) an 'indexical icon' or resemblance by association of not only the concept of writing but also writing as practice. Songs that are heard and enacted, especially for a generation coming to terms with schooled literacy, were blended into the production of Santali written texts; the act of writing linked language as text with language as sound and movement.

SCRIPT, PERFORMANCE AND THE CREATION OF AFFECT

Throughout the nineteenth century, when Santali was written, it had been written using a missionary-derived Roman script or the scripts of the dominant Indo-European languages. In the early 1930's, a Santali schoolteacher named Raghunath Murmu from a village in northern Odisha invented the Ol-Chiki script. Murmu developed the unique script for writing Santali as a way to unite Santals and other Indigenous peoples in eastern India who were spread across various states and used different scripts to write the language.

The script has different characteristics than other Indian scripts in that it is alphabetic like Roman and not syllabic like Indian scripts. Also, each sign-name indicates a picture, so for instance the letter for 't' (*ot*) is named for 'earth' and is round in shape, shaped like the Earth. All the letters in the Ol Chiki alphabet follow the same pictographic scheme (Mohapatra; Choksi, *Graphic Politics in Eastern India*).

Like the incorporation of writing (*ol*) into song to spread political messages, the creation of script was also part of an effort by Indigenous groups such as the Santals to unite their communities in order to demand a politically autonomous region within India, known as 'Jharkhand'. This movement bears some similarities to the situation in other indigenous regions throughout the world such as in North America (Bender), Southeast Asia (Smalley, et al.; Kelly, "The Art of Not Being Legible") and Africa (Kelly, "The Invention, Transmission and Evolution of Writing") in which script forms the basis for pursuing claims of sovereignty or autonomy.

In addition to the script, Murmu created an organisation called the Adibasi Socio-Educational Cultural Association in his home state of Odisha with the task of spreading the script and the message of Indigenous autonomy at the grassroots level. Migrant Santal labourers in nearby mines and steel factories supported Murmu's project and financially backed his organisation (Orans). Eventually Murmu developed printing blocks and started printing primers in the script to be distributed in villages across eastern India. The organisation Murmu founded also started schools in villages in the different regions where Santals resided in order to teach the script through non-formal literacy projects.

In addition to mobilising people through institutional means, Murmu wrote a play called *Bidhu-Chandan* to help promote the script. The play centred around two lovers, Bidhu and Chandan, who were members of two different warring clans. After falling in love in a chance encounter, the lovers become separated by war and Chandan uses a secret script—a proto-form of Ol-Chiki—to communicate secretly with her lover Bidhu. Miraculously, without any prior knowledge or training, Bidhu is able to read the script and write back. For the lovers, the script facilitated more than linguistic communication; they could use the script precisely because it allowed them to 'read each other's feelings' (Lotz 254). The lovers eventually decide to leave the Earth together and join the spirit world. Thus, while at the same time promoting traditional literacy in the script, Murmu was suggesting that, to decipher the script depended equally on the ability to apprehend commonly shared affective ties, as it did to understand the linguistic signification.

Included in the play were numerous song and dance routines that became very popular among Santals in the southern Santali-speaking regions on the Jharkhand, Odisha, and West Bengal border areas. These songs and dances, as well as the dramatic love story evoked, as one of Murmu's acquaintances told me, great *raska* in audiences, and the songs were recontextualised in a variety of settings. In the Ol-Chiki classes I attended (which were run on weekends on a volunteer basis) the curriculum often involved the learning of songs about the script. Every class began with the song that opens the play, 'Johar johar

Bidhu-Chandan ('Salutations to Bidhu-Chandan') and ended with songs from the play as well. People writing notices in Ol-Chiki would often start them with the words 'Johar Bidhu-Chandan'.

The learning of the script therefore involved familiarising oneself with a narrative and a performance modality that linked the script with feelings of *raska* as well as communication with the spirits. However, the performance of these songs also happened in the absence of this kind of Ol-Chiki instruction. In the village where I was conducting fieldwork and where there had been no informal Ol-Chiki school, local high school girls organised a nightly performance of the song 'Johar Bidhu-Chandan' in their dormitories—a ceremony in which resident non-Santal students also participated. In this setting, the prayer equates the knowledge of writing (*ol*) with good guidance and direction and establishes a communally organised affective relation with the Ol-Chiki script. While Ol-Chiki had just started to be taught in the high school by the time I arrived—and that too only in the higher grades—the prayer in the hostel, implemented by the students themselves, enacted a knowledge of Ol-Chiki beyond the limited instruction available to them. Similarly, at the boys' dormitories there was Ol-Chiki graffiti praising 'mother knowledge' and offering salutations to Bidhu and Chandan.

In addition to the play *Bidhu-Chandan,* there were popular stories circulating throughout the Santali-speaking areas that, the script was not created by Murmu but revealed to him by spirits in a dream and given to him on the top of a mountain near his native village. These stories primarily circulated through song. At a performance I went to near Murmu's native village, I saw a well-known blind, itinerant Santali musician from West Bengal and his relative, sing a popular Santali song about Murmu's discovery of the script.

> Bidu Chandan onol bonga
> wi janwar bir bonga,
> seba seba-tem konkaen, seba seba-tem konka len ...
> abowak', ol boyha abowak' roḍ.
> buru ḍhiri uḍuk' rem ṅaam keda. buru ḍhiri uḍuk rem ṅaam keda ...
> paṛsi baha teley baha hisid mey. enec' sereṅ teley lasarhed [me],
> enec'-sereṅ teley lasarhed. disom hoḍ teley atang mey ...
>
> Bidhu-Chandan, the prose-form spirits,
> the animal, and the forest spirits,
> you remained silent, you stayed in the forest ...

Our writing (*ol*), my brothers, our speech/language (*rod*), You received it from inside a mountain cave . . .
We salute you with language-flowers, we welcome you with song-dance (*enec'-seren'*)
Receive our, the people of this country's, for the sake of language, our teacher . . .

The song explicitly talks about Murmu's (the 'teacher') discovery of Ol-Chiki script in the mountain caves outside his village. It discusses the script as a fusion of writing and language, bequeathed to Murmu by the spirits including the heroes of Murmu's play (Bidhu and Chandan), the forest spirits, and the spirit of prose (*onol bonga*). The script, therefore, manifests itself in the *akhada* without any text visible; expressed through the medium of song. At the end of the passage, the participants create an indexical connection between the script, oral performance and affect, by saying that song and dance in combination with the 'language-flowers'—or sung poetry—will usher Murmu and the script into the world. This song was frequently performed in other locations too, accompanied by dance and sometimes even by spirit possession, which, among the Santals, involves a heightened emotional experience.

WRITING, SCRIPT AND THE DIGITAL *AKHADA*

Currently, the rural areas where Santals live are experiencing a surge in the use of internet technology due to the influx of cheaply available mobile phones (Choksi, "From Transcript to 'Trans-Script'"). When I had first arrived for fieldwork in 2010, many people had not even heard of e-mail or were familiar with the Internet. That has drastically changed in under a decade, as cell phone shops with cheap data plans have proliferated throughout the small markets in the countryside. Digital technology has allowed performance spaces to travel, unmoored from the face-to-face *akhadas* that were necessary to experience these heightened feelings. The spread of what could be called 'virtual akhadas' on sites such as YouTube engenders new relations between script and oral performance, such that visible graphic marks may be juxtaposed with images of song and dance in creative ways that heighten the affective resonances between them. Nowadays, promotional videos exhorting the need for writing and Ol-Chiki script appear on platforms such as YouTube, in which visual images of text and students learning

Ol-Chiki are placed side by side with dance performances and are embedded in music (https://www.youtube.com/watch?v=0pei6tse2mo). The lyrics are transcribed below:

> Ol dabon senda ge ya jati dhorom bon dohoya
> He go marsal hor te nabo nidhi bon a
>
> Let us learn our writing and preserve our tribe's religion/duty
> Hey friend, Let us follow the path of light

As you can see from the lyrics, the song is continually exhorting the listener to follow the path of 'light' through writing and in order to aid the listener on that path, images of the script and students learning the script are placed side by side with the dancing. I would argue that this is a form of language socialisation, in which the act of writing is compared with the acts of dancing and singing. *Akhada* spaces are places stereotypically associated with *raska*, and by linking Ol-Chiki with the *akhada*, clips such as these impute similar feelings to the production of written text in the script. Writing in Ol-Chiki, these songs suggest, is similar to dance and song; even though the conditions of its production may widely vary, it recalls parallel sonic and kinetic experiences within the *akhada*.

Conclusion

Ethnographies of writing have focused largely on the production and circulation of written texts. These include studies of 'grassroots literacy', letter writing, graffiti and other visible displays of linguistic information. I have also focused on the politics surrounding the production of visible marks in my own work, something which I call 'graphic politics' (Choksi, *Graphic Politics in Eastern India*). Script, I suggest in that work, is a crucial way in which Santals combat the idea among upper caste, non-Indigenous people that their language is simply a primitive regional dialect. The use of Ol-Chiki facilitates scalar links with a conception of a Santal polity that encompasses widely dispersed populations across eastern India, while the use of Eastern Brahmi connects Santals and their language to the regional territories in which they reside. This chapter shows that script emerges through sonic and kinetic embodied modalities as well, creating an ecology of affect which expands the conception of graphic politics beyond the visible.

To summarise, I argued that Santals have coupled script with oral performance practices of song and dance, crafting the script as a graphic artifact that is simultaneously charged with sonic and kinetic qualities. This could happen, as I suggested, in a political context where writing (*ol*), traditionally associated with schooling and upper caste dominant language literacy, entered into the performance space of the *akhada* as part of a movement for greater Indigenous autonomy. The invention of the Ol-Chiki script also formed a component of this political activism and its message was spread through traveling performance media such as drama, or through the work of itinerant musicians who traverse the various Santali-speaking areas. This has resulted in the script being associated with the *akhada* in emotionally relevant ways, such that the material and visible features of script recall valued and affect laden experiences of the *akhada*—experiences which now can be explicitly tied together and widely circulated through digital media such as YouTube.

As the Santali language is increasingly being taught in school—challenging the space of the *akhada* in Santali language socialisation—the question could be raised about the future role of oral performance and affective expression for the next generation of Santali speakers. Digital technology may certainly facilitate new forms of affective engagement with the language and performance traditions to provide new understandings of the idea of text and materiality. In further exploring such avenues of research, we may better understand how activists such as the publisher, who dedicates their lives to creating a future community of readers in the Santali language and Ol-Chiki script, say that while books may be important, they are only secondary to what Santals know and feel 'from the heart'.

Note

1. All translations, unless otherwise stated, are mine.

Works Cited

Banerjee, Prathama. "Culture/Politics: The Irresoluble Double-Bind of the Indian Adivasi". *Indian Historical Review*, vol. 33, no. 1, 2006, pp. 99–126.

Bender, Margaret Clelland. *Signs of Cherokee Culture: Sequoyah's Syllabary in Eastern Cherokee Life*. U of North Carolina P, 2002.

Besnier, Niko. *Literacy, Emotion, and Authority: Reading and Writing on a Polynesian Atoll*. Cambridge UP, 1995.

Carr, Gerald L., and Barbra Meek. "The Poetics of Language Revitalization: Text, Performance, and Change". *Journal of Folklore Research*, vol. 50, no. 1, 2013, pp. 191–216.

Choksi, Nishaant. "From Transcript to 'Trans-Script': Romanized Santali across Semiotic Media". *Signs and Society*, vol. 8, no. 1, 2020, pp. 62–92.

———. *Graphic Politics in Eastern India: Script and the Quest for Autonomy*. Bloomsbury Academic, 2021.

Faudree, Paja. *Singing for the Dead: The Politics of Indigenous Revival in Mexico*. Duke UP, 2013.

Hansda, Bhutnath. "OL DOBAN SERAY GIYA//New Santali traditional Song2019//Ol Daban Serai Giya Singer//BHUTNATH HANSDA", *YouTube*, uploaded by Bhasker Saren, 28 October 2019, www.youtube.com/watch?v=0pei6tse2mo.

Hembram, P. C. *Santhali, A Natural Language*. New Delhi, U. Hembram, 2002.

Kelly, Piers. "The Invention, Transmission and Evolution of Writing: Insights from the New Scripts of West Africa". *SocArXiv*, March 2017.

———. "The Art of Not Being Legible. Invented Writing Systems as Technologies of Resistance in Mainland Southeast Asia". *Terrain. Anthropologie & Sciences Humaines*, no. 70, 2018, pp. 1–21.

Lotz, Barbara. "Casting a Glorious Past: Loss and Recovery of the Ol Script". *Time in India: Concepts and Practices*, edited by Angelika Malinar, Manohar, 2018, pp. 235–263.

Mahapatra, Sitakant. *Modernization and Ritual: Identity and Change in Santal Society*. Oxford UP, 235–263.

Murmu, Ramchand. *Sadhu Ramchand Anadmala* [Poetry Collection]. West Bengal Santali Academy, 2011.

Orans, Martin. *The Santal: A Tribe in Search of a Great Tradition*. Wayne State UP, 1965.

Samuels, David William. *Putting a Song on Top of It: Expression and Identity on the San Carlos Apache Reservation*. U of Arizona P, 2004.

Smalley, William A. *Mother of Writing: The Origin and Development of a Hmong Messianic Script*. U of Chicago P, 1990.

Tagore, Rabindranath. *Shohoj Path [Easy Lessons], Vol. 2.* Vishwabharati, 1937.

Webster, Anthony. *Intimate Grammars: An Ethnography of Navajo Poetry*. U of Arizona P, 2015.

NINE

WINDING SPOOLS AND SPEAKING STONES

HOW TO PLAY BECKETT'S TAPE RECORDER AND LISTEN TO HANDKE'S STONE WOMAN IN *KRAPP'S LAST TAPE* AND *TILL DAY YOU DO PART, OR A QUESTION OF LIGHT?*

Asijit Datta

> 'What if rock refuses the stillness of being rendered a recording device, makes its own impress, exerts its own force?'
>
> (Cohen, "Posthuman Environs" 1)

The environment is a complex network of enfolded structures emanating from the interrelational entanglements between human organisms and nature. Various kinds of simulations and innate similarities are ways of decoding the affective relation between human animals and their environment. On the one hand, we detect pre-human humans borrowing from, submitting to, and imitating their terrain. On the other hand, speaking trees and animals populate the folktales and mythologies of post-language man (after structured language enters the human mode of life and living).

In Beckett's *Krapp's Last Tape* (1958), an older man unendingly listens to the voice recordings of his younger selves and examines his artistic ambitions and sacrificed love. Krapp, with his silver pocket watch, banana skins, alcoholism, constipation, black costume, and a tape recorder, is perpetually on the edges of pathos and farce. It is a technological autobiography of a sixty-nine-year-old man who is overwhelmed both by the sensual fires of his bygone days and the silence at the moment of his impending death. Akin to a critical commentary on the pauses and language employed by Krapp to dominate all the other characters in the play, Peter Handke's play *Till Day You Do Part, or A Question of Light* (2008) resurrects a stone structure. The nameless 'she' (Krapp's lover) of Beckett's play bursts into a monologue which is simultaneously an affirmation of her subjectivity and a declaration of her love for Krapp. Ironically, the stone speech also functions as an attempt to awaken the static sculpture of Krapp—the life that was sucked in by the tapes is breathed into the body by words from a primal (stone) throat. In Handke's play, we are in an environmental expanse, an open ground where the history of past humans is recorded and vocalised by stones.

The speaking stone in Handke's *Till Day You Do Part Or, A Question of Light* is a perverse environmental parody of affect, one that wrenches language from the tape recorder of the technoman and plants it back in the primal mouth of nature. However, it is intriguing to note that the stone's narrative appears from its human counterpart; the need to speak itself is propelled by a human in a text preceding the text of the stone. Thus, we are in the zone of Varela, Thompson and Rosch, who claim that,

> ... living beings and their environments stand in relation to each other through *mutual specification* or *codetermination*. Thus, what we describe as environmental regularities are not external features that have been internalized, as representationism and adaptationism both assume. Environmental regularities are the result of a conjoint history, a congruence that unfolds from a long history of codetermination ... the organism is both the subject and the object of evolution. (198–99)

MEMORIALISING AND TAPING

Krapp's Last Tape problematises the relation between memory and time. Krapp's obsession with memories is complemented and enhanced by the affect which proceeds from the inexhaustible voice of recorded selves. An affective bodily relation results from the mnemic correspondence between the disembodied voice and Krapp's (meta)physical fixation with his device. Krapp's connection with the recorder is not just fetishistic; it is one of reverence as well since the device extends the threshold of the biological body.

Beckett's magnetic tape with recording, storage and editing facilities serves as a mediating device between the old and the older Krapp, linking the past to the present's indefinite future. As a device, a tape recorder is not only an extension of the recorded voice and the person bearing it but is also defined by its simulated corporeality and plasticity. Spools bearing impressions of a previous reality are, by nature, subject to cut-ups, blow-ups and multiple other editing techniques, all contributing to a revision of reality. Beckett's Krapp himself imitates the machine recorder, and the alcohol he periodically takes, like oil, rejuvenates him. The empty stage, with its organised and compressed minimalism, resembles the concentrated inside of a recorder and Krapp, the unravelling spool.

Tape recorders have been interpreted and analysed with reference to the human body in multiple ways. Burroughs looks at language as entering the human body as external messages and the voice playback-ing it post neural processing. The left and right side of the spool are seen as the two hemispheres of the brain and have the possibilities of coupling various tapes containing sounds emanating from different parts of the body. Helen Phillips observes, '[i]f the brain's memory system is like a tape recorder, it is as if the recording head has got muddled with the playback head' (qtd. in Connor 92). Beckett's notebook reads, 'Tape-recorder companion of his solitude. Masturbatory agent' (qtd. in Connor 93), and with Krapp's 'overshooting, circling back, the tape is violently shuttled back and forth, in a sort of chronic, clonic frottage' (93). Pressing the buttons, winding, turning, looping resembles not only an act of masturbation for Krapp, but the tape itself transforms into the erotic body of the lover in the punt. Krapp is simultaneously inside and outside his recorder, with the outside Krapp waiting to insert himself inside the tape.

In his time machine, moments speak of his mother's death in autumn, his memorable equinox on a March might and his most painful memory of separation with his lover. Forty-five ghostly Krapps spread over 9 boxes, all constructed to provide temporal validation to the future Krapp, rest inside drawers to be periodically opened by the weakening Krapp. However, as ghosts they are not exorcised. Instead, as inhabitants of the house that is Krapp, they must work as bricks against the massive force of time. Ruby Cohn contends in her book *Just Play: Beckett's Theatre*, that '. . . living Krapp will never listen to what the play's title informs us is his *last* tape; there will be no more birthday post mortems. By the time of his last tape, he has sunk into a self crystallized in the soliloquy that his recording gradually becomes' (65). This irrevocably dead Krapp is precisely what Handke borrows for his stony listener. Krapp has injected his self, soul and growth into the tape to such an extent that there can never be another Krapp after death. Along with the player of the tape, the played too vanishes within the folds of history. Krapp's stillness at the end is overshadowed by the stirring tape rotating on its own. The solitary voice appearing out of Beckett's modern technological apparatus ruptures the use of conventional soliloquy on stage.

Krapp's exteriorised internal monologue from the past situates Krapp himself as the audience. And on the contrary, Krapp's present monologue converts the tape into a spectator/listener. The ritual of retrospection, introspection and repetition is undertaken to measure his current wisdom pitted against his past foolishness. Every Krapp recorded is an exact copy of the recording Krapp who is memorialising on his birthday. The last line, 'But I wouldn't want them back. Not with the fire in me now. No, I wouldn't want them back' (Beckett 28), stands in stark irony to the visual wreck that the ruined Krapp has become.

Connor underlines that 'Krapp lingers on the word "spool", the word being a veritable ideogram of the tape recorder itself, so the lovers here seem to be becalmed in a kind of pool (the palindrome of loop)' (95). Pierre Chabert's comment that Krapp is '"one body with the machine"' (qtd. in Cohn 248) comes across as all the more powerful towards the end when the unaccompanied 'tape runs on in silence' (Beckett 28). The discarded last tape encompasses Krapp's past fancies, failures, stories of sexual and bowel adjustments and his desire for the punt woman. On the other hand, the thirty-year-old tape which keeps swaddling the present Krapp within its textual folds is the one that holds the 'fire in me now' (28)—a classic technological rejection of the human. As opposed to forty-five posthuman Krapps (for all the recorded Krapps are already dead and buried inside spools) speaking from the spool, the living Krapp speaking to the spool mimics a long pause in the unceasing unfolding of nine boxes with five spools each. Krapp's stillness at the end is overshadowed by the stirring tape rotating on its own. This binarised relation is highlighted by Krapp playing the tape and the tape stopping/pausing Krapp. Borrowing Beckett's vision of self and time, Connor writes,

> The sliding of the tape is a kind of decantation, with each passing inch a diminishment of the supply tape and an equivalent accumulation on the receiving spool. At the end of the play, as Krapp listens to his younger self arrogantly celebrating his freedom from his past, the tape runs on and then, inevitably, exhausts the supply spool, suggesting the pouring out of some emblematic vessel. (95)

The negative epiphany encountering Krapp is that the onslaught of involuntary memories leaves one paralysed beyond expression. Krapp is imprisoned between his physicality trying to assert its embodied presence and his mechanised voice overwhelming him with its pastness and technological presentness. The re-production of the subject's voice is itself a displacement of the subject's existence.

If Krapp suffers from emotional, intellectual and physical constipation, then his machinistic other is the one ejecting the 'old muckball' (Beckett 24) of the past. Everything that the tape spews out (the dog, the dark nurse, the darkness, the girl) can be read as Krapp's faecal matter. Language, both as shit and as text, exceeds the being of the amnesiac Krapp. He can never entirely excrete and therein rests the problem in his character; he constructs his subjectivity around human reduction and mechanical addition. Krapp becomes the author of a tape recorder, cutting and rewinding his spools and the recorder mutates into an erotic object. Krapp's *hamartia* is in contriving his subjectivity through his past selves inscribed onto the reels as voice. The tape permits Krapp to be consumer and producer; simultaneously rewriting and erasing, the switches 'activating the powerful and paradoxical technoconceptual actors of repetition and mutation, presence and absence' (Hayles 77). If the subvocal stories (that emerge from the silent voice of the vocal folds) we tell ourselves erect our subjectivity, then Krapp's transference of those fictions onto a living machine permanently pushes him into a zone of guilt and habits marked by an obsession with time. What differentiates a personal diary from a voice recorder is the voice's hard materiality, which dispossesses the body and adds a kind of definitive truthfulness to personal history.

STONE AND MALLEABLE MOUTH

If Krapp is haunted by time, Handke's woman articulates from a zone beyond time, through and as stone. In the afterword of the book, Handke describes his play as an echo of Beckett's irreducible *Krapp's Last Tape*. He asserts that Echo in Greek mythology is 'a minor goddess or a nymph ... ("a lower rank goddess inhabiting the underwood") but definitely a woman, the voice of a woman' (Handke 103).

Associating the woman and the stone serves a dual purpose. On the one hand, it provides a medium of expression to the repressed narratives of woman and stone. On the other, the self-animated autobiography of the woman, like a palimpsest, overrides Krapp's dependent, fragmented autobiography. Handke's lover accuses Krapp of being preoccupied with material things—his ledger, pocket watch, tape recorder and the multiple Krapps which diffuse and refuse the 'original' Krapp. The greyish-white statue stutters into life to narrate the story of stone, or the woman inside stone, or the woman as stone. The text is as much an inverted *memento mori* reminding us that the dead were once living, that stony bones were flesh and speech soared from the hollow of the larynx. Simultaneously, it is a ghost story where space is claimed in the lacuna of the past through the process of continuous chronicling.

Perhaps the irony that Handke desires to press is that the past is irretrievably lost and so are the spaces associated with it. However, this irony is not the focal point of the text. Rather, Handke's emphasis is on the overflowing fullness of relational life that affixes itself to the flow of continuous life processes in other beings and things. Stone, which is unyielding and malleable, timeless yet anachronistic, acts as a representative tool for Handke to trace the unremitting movement of life. Handke's stone is both prototypical and archetypal, subject to the corrosion of time and petroglyphic obscurity. The only mythic quality the stone possesses is its plasticity against decay—one of the reasons it is often converted into gods. In the crudest sense, if human life is that which abrades and dissipates, stones too—marked by their granular decomposition into dust and their geographical dislocations—are living things. In the beginning, there was stone—human history must flow from this lithic foundation. Instead of reading human resilience and resistance in anthropomorphic terms, they should instead be understood as borrowings which are petromorphic in nature. Stone is not deathless; it simply bears stories of destruction, tectonic convulsion and colonisation. It is essential to remember Jeffrey Cohen's *Stone: An Ecology of the Inhuman*, where he writes that,

> ...we map our progression as a move from the Paleolithic to the Neolithic, charting cultural development by increased intimacy to stone. Yet *human* immediately becomes *posthuman* as a consequence of the enlarged temporal frame that geology demands. Such a stone-etched countervision invites reflection on what it means to inhabit a world that is at once potentially indifferent to humanity yet perilously continuous. (59–60)

What Handke's woman recounts is both personal history and the history of interdependent existences. Two bodies carved out of stone and erected as monuments of memory are identified first on the grounds of gender and then on the basis of surplus life that emanates out of the Romanesque woman. Krapp's graven image is 'dead and gone' (Handke 4) and as if by a tragicomic aberration of time 'a drunk on his way home, has painted lips a deep red, stuck a red cardboard nose on him and wrapped stuff that looks like bits of tape from a cassette round his forehead' (4). Built on the same stone platform, the woman stands physically detached from Krapp, exuding hallucinatory, dreamlike motion from her 'whole face' (Handke 4) and 'half-naked ... ample bosom' (5). Standing vertical beside Krapp's stony immovability, the unstoppable speaking woman rails at Krapp's self-devised pauses for effect, his 'liturgies of pauses ... psalms of pauses' (9). Upholding her speech as signalless sign, rather than a quest for meaning, the stony woman speaks and acts 'in my own language ... the language of my childhood and the language of my senses' (7).

In *What is Philosophy*, Deleuze and Guattari, while discussing the independence of art works, mention that art preserves 'a bloc of sensations ... a compound of percepts and affects' (164). Existing autonomously outside its creator, viewer and hearer, a work of art 'could be said to exist in the absence of man because man, as he is caught in stone, on the canvas, or by words, is himself a compound of percepts and affects. The work of art is a being of sensation and nothing else: it exists in itself' (164). Krapp, the artist of both his future and the subjects of his tape, is negated by the monument that opens its mouth—'A monument does not commemorate or celebrate something that happened but confides to the ear of the future the persistent sensations that embody the event: the constantly renewed suffering of men and women, their re-created protestations, their constantly resumed struggle' (177).

Krapp's consort emerges as a stone that 'opened its eyes' (Handke 5) to protest against Krapp's time-bound philosophy of silence. Ironically, the living Krapp is accused of being impermeable, 'not enough of a shaper ... a creator' (12), a reference to the conventional reading of stones as inert and impenetrable. The unknown woman deconstructs Krapp's 'Temple of Neverending Interpretation and Signification' (18), his consciousness perpetually trapped in the 'something behind the something, the idea of something' (17), his random pointing everywhere akin to a child and his awareness of his mother's womb as 'a sepulchre' (19). Krapp 'incapable of a duologue' (20), along with his validating tape, philosophises life as a congregation of moments. Therefore, anyone who insists on augmentation of these moments is either rejected or sealed off and locked in the past via the tape recorder. The woman in the punt is doubly subordinated in Krapp's last heard tape; like a dramatic monologue, Krapp assumes the role of a 'solo performer' (21) and the woman is silenced and acted upon; on the other hand, the history and name of the woman who decentred him from his 'joie de vivre' (24) are erased. The re-sounding echo from the resounding woman of the tape not only overwhelms the spools, which are human dependent instruments, but crack open the dead Krapp, who 'appears to be becoming like her, even if almost imperceptibly' (32). The last tape that Krapp encounters is the organic, sentient, wriggling tongue of the stone woman.

Against the tape recorder of Krapp, Handke's ancient stone is more an agent of liberated embodiment—a specific kind of freedom that is experienced in entanglement. Therefore, the speaking stone (an immersed voice), and not the recorded technological voice (belonging to the individual), is charged with an artwork's universal appeal. Human texts (presented in the case of Krapp as the fixed recorded voice from the past) are permanently embedded with a predetermined hierarchical centrality and deep-rooted anthropocentrism. Stone narratives disempower humans and remove them from the linear movement of life, death and historical truth. Perhaps, this is why Handke lets his stone begin with the words, 'My act now. Your act's over, Herr Krapp' (6) as if to remind us that the (pre)historic is the only modern and it begins its recording with the death of human time as we know it. If human time is marked by subjectivity and temporality, then stone time is objective and transcendental. Cohen holds that 'Creating art with stone is not the domestication of an element, but a human-lithic collaboration that recognizes the art stone already holds' (*Stone* 61).

After exhumation, bones resemble stones; both evolve from putrefaction and are relics of the savage and the civilised. However, there is an essential difference between stone and bone. If stones are employed in pre-history writing (such as cave etchings and paintings), Stonehenge, statues, mansions and even wars, the skeleton, in the popular imagination, is either a medical prop or a ghost of vengeance. Human love for stone emanates from the inhuman endurance that inheres in stone—the human evaporates inside stone time. It is difficult to unravel whether in Handke the stone activates the text, or the text touches the stone. What is certain is that against Krapp's limited acts of rewinding and forwarding the tape, the stone inserts a non-native, terrestrial duration within the text. All we can approximate is that the woman is closer to bone and Krapp affixed to the flesh.

Krapp keeps himself occupied with the calculable passage of time, with remembered beginnings and predicted endings; his obsession with restricted historical frames is borne out by his use of the recorder in a struggle to sustain himself as the prime mover against a non-anthropocentric progression of time. To counter the ghost of the nameless woman 'among the flags' (Beckett 27), Krapp records the living presence of Effie, Fanny and Vespers in his last tape. Living Krapp's last line, 'Lie down across her' (B27), can be read as the opening of a grave and a falling into death. Handke disinters this grave of the lovers and listens to the ceaseless disquisition of the lover's ghost. Over Krapp's documented archives, the stone woman's non-textual, lithic expansiveness assumes a heterogeneous dimension. Handke's stone history, though personal, expose insecurities about Krapp's non-written, a-textual, storied matter.

On stone time, Cohen remarks that 'Stone challenges small, linear divisions of human history through its aeonic insistence. The lithic thickens time into multiple, densely sedimented and combustively coincident temporalities' (*Stone* 78). Handke, akin to an archaeologist, unearths the epochal stories embedded within the stone, 'the long transmission of the messages . . . ruins, petrified remnants, chips of flint speak the narratives of those who once touched their hard surfaces' (Cohen, *Stone* 103).

The stone endures and the reel rolls. The tape, through its inorganic organisation, affects the responses of Krapp's body. Solidarity between Krapp and his tape, based on mutual affecting and manoeuvring, stands as compensation for the failure of the aesthetic and the emotional Krapps. Ironically, the inorganic affect that is generated by the device is carried through the organic, gesticulating voice of Krapp himself, which in turn is evidence of previously affective states. The tragedy of Krapp is primarily in his deliberate negation of the material thresholds of the recorder, 'the limits that determine the potential affects an object can generate, which in turn define what that object is' (Ash 3). Repeated rewinding and fast-forwarding are efforts to alter the material thresholds for an enhanced ontological affect; perhaps, Krapp's posthuman desire is for the tape to assume personhood, to become the punt-woman, to overwhelm him with organic affects rather than inorganic ones. As his only companion, the tape has, over the years, altered the material and physical components of Krapp's body—Krapp is an old alcoholic with a strenuous walk, near sighted, hard of hearing, constipated and one who embraces death at the age of sixty-nine. Every year, the ritualistic realisations of failure that appear from the tape shrink Krapp and speed up his aging mechanics. This is the affective afterlife that James Ash talks about while explaining tinnitus (5). The autonomy of the tape reminds one of the contributions of Bernard Stiegler, who is 'most concerned with how various forms of inorganic organised matter operate as forms of tertiary or externalised memory and how that memory shapes humans' primary retention (perception) and secondary retention (memory)' (Ash 7). Krapp anthropomorphises the machine and transforms humans into faltering memories and vocal remainders. Even Krapp's movements parody the tape's stopping, winding and switching off. The posthuman technological other of Krapp exhausts and dissipates the authorial mind itself—the only thing which remains on the darkened stage is the still light of the tape gone silent.

Arnold Berleant observes that no other inorganic object possesses a history older than stone which was used as building materials for 'monuments . . . for the dead . . . statues and gravestones . . . sealing the body in a sarcophagus, stone crypt, dolmen, mausoleum, or pyramid' (52). Similarly, Rosi Braidotti puts forward her argument for a 'post-anthropocentric posthumanism', a kind of 'posthuman ecocriticism' that listens to 'animal, water, stone, forest and world' as acting and 'story-producing' entities (Cohen, "Posthuman Environs" 7). Simultaneously, Cohen stresses that the Disanthropocene is the impression of non-organic natural elements on our bodies, stories and worlds (24). Handke's drama borrows the form and medium of stone for the narrative of the woman. The stone's impression upon the woman's body is so concentrated and dense that only the ecological and indigenous cognition belonging to the stone is deployed to counter Krapp's technological one. Handke ends the play with the affect exhaling from the stone woman altering the tombstone of Krapp, making it more alive—'. . . the male figure beside her appears to be becoming like her, even if almost imperceptibly' (Handke 32). Beckett's living bodies are first turned into sculptures in Handke and later—out of an affective ecological mechanism—the stones unfurl their onomatopoeic history of flesh and foundation.

Works Cited

Ash, James. "Technology and Affect: Towards a Theory of Inorganically Organised Objects". *Emotion, Space and Society*, vol. 14, 2015, pp. 84–90.

Beckett, Samuel. *Krapp's Last Tape and Other Dramatic Pieces*. Grove Press, 1978.

Berleant, Arnold. "The Soft Side of Stone: Notes for a Phenomenology of Stone". *Environmental Philosophy*, vol. 4, nos. 1 & 2, 2007, pp. 49–58, https://www.jstor.org/stable/10.2307/26167140. Accessed 8 Aug 2019.

Cohen, Jeffrey Jerome. *Stone: An Ecology of the Inhuman*. U of Minnesota P, 2015.

———. "Posthuman Environs". *Approaching Posthumanism and the Posthuman*, June. 2015, U of Geneva. Plenary Address.

Cohn, Ruby. *Just Play: Beckett's Theater*. Princeton UP, 1980.

Connor, Steven. *Beckett, Modernism and the Material Imagination*. Cambridge UP, 2014.

Deleuze, Gilles, and Félix Guattari. *What is Philosophy?*. Translated by Hugh Tomlinson and Graham Burchell, Columbia UP, 1994.

Handke, Peter. *Till Day You Do Part Or, A Question of Light*. Translated by Mike Mitchell, Seagull Books, 2019.

Varela, Francisco J., et al. *The Embodied Mind: Cognitive Mind and Human Experience*. MIT P, 1993.

Hayles, N. Katherine. "Voices out of Bodies, Bodies out of Voices: Audiotape and the Production of Subjectivity". *Sound States: Innovative Poetics and Acoustical Technologies*, U of North Carolina P, 1997, pp. 74–96.

TEN

EMBODYING THE *BIRANGONA*

NEGOTIATING HISTORY AND MEMORY OF THE BANGLADESH LIBERATION WAR, 1971

Antara Ghatak

Although history has been one of the main disciplines through which we can understand gender, it has been contested for being situated in a particular time and universal in delineating experiences. With the aim of 'historicizing the discipline', Danish social anthropologist Kirsten Hastrup argues that the apparent unity of history as linear and causative is a discursive rather than a social fact, since analysis reveals the non-synchronicity and discontinuity of social experience (1). With a set of reflections 'upon the nature of history', *Other Histories* edited by Hastrup demonstrates the polyglot nature of history. The inclusion of anthropology in the study of history has led to rewriting history. Formerly, only records provided acceptable historical evidence, but from the perspective of anthropology recollections are equally valid, given that in both, culture is significant. According to Hastrup, historical anthropology seeks to explore the 'state' which is obtained in a particular time–space.

Over the last quarter of a century, historians have been engaged in the process of rescuing women of the past from their invisibility in traditional records. This work of recovery and retrieval has made possible the redefinition of 'history' to include aspects of life previously seen as 'non-historical'; they were natural and therefore timeless and unchanging, especially in the context of family relations, domesticity, and sexuality. The importance of reconstructing women's history by women is clear enough. It allows for the retrieval of life stories as conceptualised and told by them—real people, real lives as opposed to the monopolised historical accounts of gendered experiences. With the absence of voices in gendered versions of history, memory becomes a tool through which the lived experiences obliterated from mainstream discourses could be restored. To understand the complexity of 'modern' histories and examine objective conditions which are matched by a subjective reality, the intersection of history and memory becomes central to the study of historical anthropology.

Feminist scholarship has clearly shown the interconnections between history, memory and gender. Marianne Hirsch and Valerie Smith in 'Feminism and Cultural Memory: An Introduction' note:

> For the last thirty years, feminist scholarship has been driven by the desire to redefine culture from the perspective of women through the retrieval and inclusion of women's work, stories and artifacts. This period has also seen an explosion of literary and cultural production by women in numerous languages and cultures that in itself has shaped much of cultural memory of the late twentieth century. (3)

When viewed through the lens of gender, history and memory reveal the complex construction of the body and its relationship with the semantic structures of society, religion, culture, economy and ecology. The body becomes a nuanced site as on the one hand its meaning is extracted from the realm of the physical object where it is coded with political meanings, and on the other it demands to be a human subject decoded by lived experiences. Moreover, notions of 'embodiment' and 'expressivity' lie at the heart of feminism and when this is raised in the context of history, memory and gender in a post-conflict society, one walks a tightrope.

In a post-conflict society, women have often been blended into the traumatised landscape and thereby placed in the realm of the 'objective'—quantified in terms of data and facts. The woman's body in a violently altered environment requires critical revision as study brings to light the complex nuances regarding embodiment and expression. We need to explore the 'subjective' experiences of women separately, as the distinctive characters of their bodies define their human existence in their immediate environment. Such experiences form a series of self-enclosed fragments having their own integrity and the potential to attain meaning on their own terms, without having to place them within larger historical narratives and processes.

The nine-month Bangladesh Liberation War disrupted the ordinary lives of women, changed the ritualistic structure and continuity of regular life and altered their naturalised behaviour. It becomes important to note here that even in the post-war society of Bangladesh, the newly formed polity and successive governments refused to acknowledge the experiences and narratives of ordinary women by reducing them to aggregated groups who had no role to play in the extraordinary events of the war. The atrocities faced by women, their trauma and loss, were repressed by official historiography while glorifying the war, and in doing so it denied women as active agents in crafting the new social transformation.

The argument in this paper is driven by an engagement with the narrative of victims and survivors of rape and their sufferings because of essentially bearing the burden of gender and biological inequity. Sexually violated women of the war were gloriously invoked by Sheikh Mujibur Rahman as the 'birangonas' or 'war heroines' but they eventually became the rejected and marginalised population of the historical fallouts of the Bangladesh Liberation War. Due to the limited scope of this paper, I will refer to the experiences of Ferdousi Priyabhashini (1947–2018) the *birangona*, one of the first women to testify about her victimisation in 1999 and explicitly talk about her rape. My objective has been to show how Ferdousi, though victim and sufferer, is also a protagonist who told her stories to the world defying the politics of embodiment and expression. Denied a place in a logocentric historiography, she sought refuge to an alternative mode to express her immeasurable sorrow and indomitable spirit so that her narrative could be used to memorialise and celebrate a human legacy, and carry forward a message of perseverance.

A closer scrutiny of the historiography of the Liberation War is required if we are to understand how it incorporated gender-specific narratives that conditioned us to remember some events at the expense of others. The historiography of the Bangladesh Liberation War, though full of its own kind of righteous and humanist glory, generated a grand narrative. It served to ameliorate the brutality of the crimes committed on women during the period of nine months of the war and thereafter excluded them from mainstream discourse. The memories contained in the narratives written by Ferdousi on the war serve to debunk the cultural politics associated with the body and representation. Her narrative does so by providing internalised accounts of trauma and pain, which bring into focus complexities of history, memory, and gender.

BIRANGONAS OF THE BANGLADESH LIBERATION WAR

In August 1947, British India gave rise to two new states, the Dominion of India and the Dominion of Pakistan. The Dominion of Pakistan comprised two geographically and culturally separate areas to the east and the west of India. The western zone, consisting of Urdu speaking Muslims, was popularly termed West Pakistan and the eastern zone, consisting of a Bengali speaking population, was initially termed East Bengal and later, East Pakistan. Ayub Khan, the first military dictator of West Pakistan, highlighted the ethnic, linguistic and sectarian nature of these two states, and aimed to reform the East Pakistani, Hindu-like Bengalis into 'good Muslim' subjects (Saikia, "Ayub Khan and Modern Islam" 296). This ongoing ideological divide, along with political and economic discord between these two parts, led to fragmentation. In 1971 a civil war was fought between West and East Pakistan and an international war between West Pakistan and India. These two wars in 1971 are generally referred to by a single name: the Bangladesh Liberation War (*Muktijuddho*). The armed conflict started on 26 March 1971. The war ended on 16 December 1971 and resulted in the secession of East Pakistan, which became the independent nation of Bangladesh. Officially, 3 million people were killed and 30 million became homeless (Debnath 49). The war, with the aim of ethnic cleansing, perpetrated a reactionary genocide and mass rape, leaving behind millions of suffering peoples.

At the end of the Liberation War, Bangladesh was left with a staggering number of victims: 200,000 to 400,000 women raped and 3 million killed (Debnath 46); between 25,000 and 195,000 women were said to have become pregnant (Mookherjee, *The Spectral Wound* 325); 30,000 had committed suicide; 3000 war babies were born (327). Thousands of Muslim women and Hindu refugee women disappeared leaving no records. Women and girls from the ages of seven to seventy-five were raped, gangraped, and either killed or taken away by the military to become sex slaves to officers and soldiers for the duration of the war. It should be mentioned here that many women were not only raped but were also often forced into sexual slavery. Their bodies were ripped, their sexual organs mutilated and bayoneted, and the naked dead bodies used to feed scavengers (Hassan 29).

The war ended on 16 December 1971 and as early as 22 December A.H.M. Kamruzzamman, Home Minister of the interim government of Bangladesh, announced that all young girls and women who had been subjected to inhuman torture by the occupying Pakistani army in the last nine months would be accorded full respect as *birangonas* of the Bangladesh Liberation struggle (Mookherjee, *The Spectral Wound* 129). Sheikh Mujibur Rahman, the first Prime Minister, on behalf of the newly liberated state of Bangladesh declared that the women who had been sexually tortured during the war by the Pakistani army were war heroines.

Birangona is a Bengali word which means 'brave heart women', but this was a specific title conferred on them. It is important to note that the *birangonas* were distinctly demarcated from the *muktijoddhas* or freedom fighters (a term that specifically refers to Bangladeshi males who fought in the war). Moreover, the bodies of the *birangonas* became the site where power relations were meted out by the warring nations. For the Pakistanis, the rape of women was a means of showcasing their military prowess. For the Bangladeshis it was a matter of lost pride, for women were venerated as 'mother' by conservative Bengali society.

In many ways, the history of the Bangladesh Liberation War was concerned with addressing certain fundamental questions which expressed a need to rationalise and periodise the events of 1971: the nature of the Liberation War; the political and military organisation; the attack on the intelligentsia; and the encroachment on human rights and liberties were some primary concerns. The lived experiences of women were never heard or documented by the official discourse, thereby failing to acknowledge their sufferings and sorrow during and after the war. Even after the passage of fifty years, the warring nations continue to maintain an uneasy, volatile, brittle peace.

The collective memory of post-war Bangladesh reduced the *birangonas* to imaginary and symbolic entities. Their rape was quantified in terms of ethnicity, religion and nationality. The stories of rape were drawn out of the province of male desire of domination and towards an assertion of power and politics. Cultural narratives thus became fundamental to social and political life as a means of understanding the cause of rape rather than the effect. The middle class and political elites sought to focus on the romantic, nostalgic picturisation of Mother Bengal ravaged by enemies, her honour tainted and her purity defiled. The official accounts of wartime rape and violence explicate the sexual and ethnocultural mastery of the Pakistani army. Instead of expressing the painful and personal violation of rape, popular representations of the nation as a venerated mother entered public circulation through media and literature arousing sentiments of nationalism and pity. The cultural fantasy surrounding the figure of the raped woman thus contradicted the realness of their experience and defined specific gender roles and categories for them. They were rendered as passive agents who were waiting to be delivered by the philanthropic policies of the state.

In the aftermath of the 1971 Bangladesh War, memory plays a crucial role in the search of representation or self-vindication for the women. The reconstruction of past events causes a piecing together of fragments of memory with the hope of understanding, remembering and reporting the violence—to move on. With the rapid passage of time, memory remains the only possible means of self-expression, which in no way qualifies as authoritative evidence but leads to the production of new knowledge. The writing of experiences of rape and trauma can be a call to defy foundational histories which essentialise explanations and classifications.The emergence of women's personal narratives have allowed the inclusion of new voices. The memoirs, autobiographies, testimonials, and fictional narratives written by women about this period for the first time challenge the official version which had essentialised the *birangonas* as physical bodies while negating the profound emotional aspects of their lived experiences. These personal experiences escape the grip of a masculine cultural ideology and lay the ground for the development of a critical feminist scholarship in post-war Bangladesh. The experience of war crime or rape which is foregrounded through memory attempts to see beyond the physical act to the social, and finally to the psychological. In this context, the memories of women written during the 1971 Bangladesh War serve the purpose of preserving the memory–identity of those otherwise lost to an audience too eager to frame larger public and political discourses.

The two empirical components that I have referred to here are the memories of Ferdousi Priyabhashini published in Yasmin Saikia's seminal book *Women, War and the Making of Bangladesh: Remembering 1971* (2011) and a content analysis of *Bangladesher swadhinata juddho dalilpatra* (*Liberation war documents*) published by the People's Republic of Bangladesh, Information Ministry, from 1982 to 1985. An analysis of Ferdousi's memories, when set against the official historiography, illuminates the nature and extent of the divergence between women's own remembered accounts and their public construction as both victims and agents in the war.

Moreover, the study of the personal memories of Ferdousi becomes problematic as she defies the logocentric practices of power and seeks to find expression in the natural objects around her. She created her sculptures out of dead and abandoned barks of trees. In her testimonies till her death in 2018 she had stated that the dead wood of trees is symbolic of her body which was physically and socially debilitated by the virtue of being a *birangona*. Objects drawn from the natural world helped her find resolution to the affect of the war that she had embodied in herself. I aim to discuss how the memories contained in the narratives and sculptures of the *birangona* Ferdousi Priyabhashini challenge the gendered nuances of the Bangladesh Liberation War of 1971.

HISTORICISING THE BANGLADESH LIBERATION WAR

The fifteen volume *Bangladesher swadhinata juddho dalilpatra* (*Liberation war documents*, 1982–1985) was the first official historiography of the nation. The 'Preface' mentions, 'The primary task has been to collect and publish documents. The main consideration was to have correct documents for correct events' (v). Moreover, there is mention of a collection of more than three and a half lakh documents which consisted of published gazettes; procedures of the parliament; reports and verdicts of court cases; reports of commissions; activities and proposals of political parties; proposals of public rallies and meetings; various students' protests; media reports; data and documents provided by acclaimed persons; political documents; and the governments' proposals and directions. Responses from the international community; private diaries; letters; memoirs; interviews; official documents pertaining to administrative control by the government and the freedom fighters over the areas freed from the Pakistani army; documents regarding war strategies; and the documents pertaining to the participation of the civilians in the war form the bulk in the project. The primary sources of these documents as mentioned in the 'Preface' are libraries, educational institutions, newspaper offices and the questionnaires framed by experts that were distributed among common people.

At the same time, the editor Hasan Hafizur Rahman says that individuals from the entire cross section of society had participated in the Bangladesh Liberation War and that a comprehensive narrative would be incomplete without their stories of the myriad sacrifices they had made and tortures they had endured. It is only the eighth volume of *Bangladesher swadhinata juddho dalilpatra* that focuses on the genocide and the sufferings of women. In form of testimonies from the rape victims themselves, eyewitnesses and bystanders, they explicate the torture perpetuated on women by the Pakistani army.

The narratives of women contained in the eighth volume were not adequately proportional to the military, political and nationalistic discourses that occur in the fifteen-volume book. Moreover, although such narratives bring out intensely personal and episodic events, these somehow do not include the comprehensive continuum of pain of the ordinary women associated with the war. These narratives of war were drawn from the domain of published and archived records. History has generated statistical data pertaining to women through various official procedures. The absence of their pains, struggles and sacrifices is starkly visible in any such survey. The experiences of women were overlooked by historians and researchers as these experiences were deemed to be 'non-historical'.[1] Recent scholarship on the Bangladesh Liberation War points out the inadequacy of representation of women during the war. Ranabir Samaddar comments on the absence of an 'authoritative and comprehensive history' and remarks that 'a comprehensive history thus has to rely ironically on fragmentary experiences. The question of historicizing 1971 lies in those experiences' (221). He specifically mentions that the experiences of women form a huge body of narratives that need to be included in historiography.[2] Yasmin Saikia, Nayanika Mookherjee and Bina D'Costa in their study of the representation and memorialisation of women in the Bangladesh Liberation War have pointed out the serious omission of women's experiences in national narratives. On being asked about the representation of women victims in Bangladesh, Tariq Ali (one of the trustees of the Liberation War Museum) commented that there were over 40,000 pieces of writing in their archive which talk about how the birangonas committed suicide, were killed after being raped, or had vanished into India. However, he claimed that that there was nothing that commemorated the sacrifice of these women because Bangladesh is a male-dominated society.[3]

Such an exclusivist approach had resulted in certain state policies undertaken by successive governments in Bangladesh. It is important to note that *birangonas* never received the status of the freedom fighters. On 15 December 1973, the government of Bangladesh conferred the title of 'Bir Protik', the fourth highest gallantry award in the country, on 426 freedom fighters. Only two women were included in the list—Dr Capt. (Retd.) Sitara Begum and Taramon Bibi. Five decades later, the High Court of Bangladesh ordered on January 27 2014, that the *birangonas* should be recognised as freedom fighters. It was only in October 2015 the People's Republic of Bangladesh published a government gazette containing a list of 41 *birangonas* who received the status of *muktijoddhas* ('freedom fighters') and were therefore entitled to all the associated facilities and benefits (D'Costa 78). The Dhaka Tribune reported on 13 October 2015 that Liberation War Affairs Minister, A.K.M.

Mozammel Huq, declared that the allowances for a freedom fighter aged over 65 would be Tk 10,000 (103).[4] The government has also announced that this list of female freedom fighters would be updated at regular intervals in the future. On analysing the available data, we can infer that the representation of women was selective and constructed by a widely circulated discourse—that is, mainly official data and media representation—which in turn was built and governed by the state policies. These discourses negated the experiential dimensions of the war and located women under the gaze of an essentially patriarchal system which marginalised women in the entire corpus of the war narrative.

Ferdousi Priyabhashini was one of the *birangonas* who have managed to tell her stories to the world. She was the first Bangladeshi woman to testify to her rape by the Pakistani army during the war and her subsequent rejection by her own society. In an interview with Azra Rashid, Ferdousi asserts her status as *birangona*:

> I saw the war, bore the brunt. Maybe I didn't take up the gun and fight, but I fought with that time. Is it not part of the war to live in the concentration camp? I have been raped. Was that not part of the liberation war? I don't feel comfortable to become freedom fighter because I did not fight with a weapon or anything. I think I am a rape victim. A *birangona*. (148)

Her first-person narrative can be found in Yasmin Saikia's seminal book *Women, War and the Making of Bangladesh: Remembering 1971*. She tells us about her life before the war, narrates accounts of rape and other acts of terror and torture perpetuated on her during the conflict, and her trials in post-conflict times. It is here that she goes beyond the nationalistic discourse of the aggressor and the victim to a deeper awareness of how violence continues to haunt women's lives even after the end of a conflict and initiates a conversation with memory and memorialisation while travelling through the difficult and varied terrains that women's political struggles and movements have taken in Bangladesh post-1971. Ferdousi's life story attempts to integrate stories of gendered violence into the corpus of national history by removing the veil of silence surrounding their experience.

In the section titled 'Victims' Memories' in Saikia's book, Ferdousi narrates incidents from her life prior to her rape in the war. In doing so she conveys not only her personal suffering, but also the pains and sorrows of thousands of women. Her narrative could be seen as essentially trapped in the title of *birangona* and her experience of rape. But by narrating her story in totality—that is before and after the war—she tells us that the ultimate cause of her suffering lies in the patriarchal structures and practices of society. Having described the span of the nine-month war as a 'nightmare', she talks explicitly about her ostracisation by her own family and immediate society:

> After 1971, I was considered as a prostitute in Bangladesh, at least by my family. I was not allowed to attend wedding ceremonies as I was considered inauspicious ... The first time I came out publicly about my rape after 1971 was in 1999. It took me twenty years to regain some sense of self-respect and be able to talk about it. (Saikia, *Women, War and the Making of Bangladesh* 165)

In the aftermath of the Bangladesh Liberation War, the raped woman located her internal and social agency in relation to the gaze of the man whose eyes reflected those religious ideals that paradoxically produced her as subject and commodity, and whose gaze was entirely regulated by patriarchal social systems. The war-ravaged women were summarily homogenised as bodies either violated by the Pakistani army or waiting to be delivered by a benevolent state but destined to witness ruthless rejection from post-war society. Their multiple identities, experiences and wounds were negated under the label of *birangona* which was the sexually violated body.

Sheikh Mujibur Rahman, the first Prime Minister of liberated Bangladesh and referred to as the 'Father of the Nation', assumed a paternal role for those women who had been sexually violated during the war. His government undertook several programmes to reinstate the raped victims into society. Centres were set up in various districts of Bangladesh to provide women with training in areas such as sewing and nursing, to help them earn a livelihood. The state initiated a 'marry off campaign' which would enable women to get reinstated into domesticity. Abortion and adoption programmes were also launched by the government. Unfortunately, these projects were abandoned with the assassination of Sheikh Mujibur Rahman in 1975.

Accounts of wartime rape and violence explicated the sexual and ethnocultural mastery of the Pakistani army over the bodies of women, while Bangladesh sought to focus on the romantic nostalgic picturisation of Mother Bengal ravaged by enemies, her honour tainted and her purity defiled. But the truth is that conservative society did not accept these women back into its ambit. In course of time stories of rape, as they were drawn out of the province of male desire and domination towards a realisation of the notions of body and power, were lost. The state which positioned itself as a paternal figure could deliver only a symbolic performance. *Birangonas* trapped in their titles were left behind.

> I had told them then and now I will tell you now that what I regret most is what happened to women after 1971. I cry for our lives post-1971. We were physically assaulted in 1971, but after 1971 we were both physically and mentally assaulted. Today I have no shame to say I am a victim. But I continue to ask why are only the victims identified? Where are their rapists? (Saikia, *Women, War and the Making of Bangladesh* 166)

Memories contained in abstraction were only means through which Ferdousi could share her lived experiences. The recollection of the past in retrospection was her only means of restoring her voice in the historiography of the state. She was the first to lay bare the fact that the *birangona's* story is rarely integrated with the corpus of national war narratives. The *birangona* operates more as a declarative term rather than an acknowledgement of their individual agency during the course of the war. The role of *birangona's* own voice in the construction of these radical identities is thus crucial. She breaks through her own and thousands of other women's omission in official historiography and reveals the complex nuances that are associated with gender and representation.

Embodied in the title *birangona*, Ferdousi lacked linguistic agency and failed to learn the language and mechanisms of patriarchal power to transcend the social stigma of being raped. Having been denied a place in history and society she scrambled to express and speak for herself. The representation of the *birangona* through official discourse was not simply about giving them voice, but is also concerned with constituting, working for and representing the positional relations of the woman to the dominant and *vice versa* within a cultural context. By the act of rape, Ferdousi's body had come under the influence of an overarching state policy which expected a degree of behavioural policing along with societal demands that urged silencing the traumatic experience of rape. The image of the *birangona* that had emerged through collective discourse represented the raped woman as unbalanced, as physically and psychologically debilitated. Ferdousi's testimony in this respect calls for a reappropriation of control and voice by reasserting her power of self-representation. The representational problem of speaking for the 'other' is based on the notion that the subject's social location and position have significant impact on how her speech is received. Discourse analysis reveals that 'who' says something can drastically affect how it gets heard. The state legitimises its function as the privileged speaker speaking on behalf of the less privileged, thus serving the purpose of perpetuating existing power relations between itself and its citizens. By re-appropriating her narrative, the victim regains discursive power and establishes her unique position as her own ethnographer. This allows her to achieve a level of representative control not afforded to the female subject in state testimonies and documents.

Significantly, Ferdousi chooses art as an expression of memories of rape. Through the constant negation of the society, she realised that her voice would never be heard or understood since it lies beyond the political and cultural logic used to justify the events of 1971. She worked with a specific kind of 'found object': dried roots, tree barks, and other tree waste discarded on the city roads by timber merchants and other capitalist enterprises. In an article by Nayanika Mookherjee, published in the *Daily Star* on 15 March 2018, Ferdousi explicates the reasons behind her choice to sculpt out of discarded matter. She states, 'I work with and sculpt the bark of trees as they signify to me the abandoned, like I was after 1971—abandoned, ridiculed, humiliated and treated with contempt' (Mookherjee, "Lessons of Consent and Critique"). This attempt to create significance out of waste becomes the metaphor of her life. The act of seeking out, picking up and transforming the debris of the everyday enacts the retrieval and revival of the violated self. In defining her creative philosophy, she discards the label of an artist and grounds her praxis in everyday work. She resists the overarching patriarchal hegemony that discards the raped woman and eventually assumes the role of their saviour. She adorns herself by her own means and generates her identity through autonomous resources.

Sorrow, numbness, shame or guilt often function to distance the subject from the pain of events, offering a form of safety in unspoken utterance. These emotions facilitate a process wherein the subject covertly returns to a world of the creative imagination. The real nature of the past event, which is recollected by the victim, is often complicated by the question of authenticity. For Ferdousi it was the negation of expression through words. She refused to be imprisoned in the material effects of the catastrophe and her narrative moved beyond the immediacy of the event towards an understanding of corporeality, identity and discursive power in the fields of pain, struggle, and healing through the discarded objects of nature. As memory may exist as raw perceptual moments, the individual creative potential which shaped the abandoned objects into tangible forms associates the past and present through a series of symbolic images and interpretive patterns.

Ferdousi's sculptures disembody cultural practices of representing rape and critique the patriarchal discourse of embodying the raped body with specific markers of disgrace, ruin, vulnerability and loss of agency in the context of the Bangladesh Liberation War. The artwork dismantles social and lexical modes of expression in a manner similar to how she condemns the popular assumption of the *birangona* as a woman without honour or repute, waiting to be delivered by the state. Through the process of devaluation of the linear pattern of history, Ferdousi opens up the possibility of an alternate history which promotes a dissociation from the hegemony. She realised that the materiality of violence and trauma is difficult to express in words and is often shrouded by textual silences which render the victims as passive and helpless. Moreover, the lexical representation of rape is strewn with words like 'dishonour', 'guilt', and 'shame', which have tied the *birangonas* down. To defy the male rhetoric surrounding rape, she tries to seek to find a language that would enable her to find a definite identity in the future. Her embodiment of the title of *birangona* reiterates these crises of expression and therefore reminds us not only of the limits of narratability, but also the sociological, political and material forces which function as sources of resistance towards the rhetoric of rape and violence. Her sculptures of dead wood and abandoned city matter reflects her resilience to adapt, engage with and shape her experiences, and creates the opportunity to find a 'third space' where the political potential and communal identity of those experiences may be explored.

As a survivor, sculptor, social activist and above all a *birangona*, Ferdousi not only implicitly demands more stringent laws and greater state intervention in the rehabilitation of *birangonas*, but her work also highlights the appalling lack of will demonstrated by the state and the possibility that the state would abuse the greater powers that accrue to it. The figure of the *birangona* as victim has been invaluable to gendered representations of the 1971 war. In pointing to the systematic character of gender domination, Ferdousi Priyabhashini brings out the notion of women's oppression as a multifaceted and contradictory social process during the Liberation War and its aftermath. She attempts at pushing the limits of women's expressions in Bangladesh. As she scrambles to learn the language and mechanisms of patriarchy to express her insurmountable pain, she realises the impossibility of the task. She finds her own medium of communication and renounces the privilege of using words to express that which she had been hitherto denied. She reminds us of the myriad non-linguistic ways of expression and probes us to reflect and retrospect on the cultural politics surrounding body, affect and the war. As a raped woman she refuses to submit her internal and social agency to the gaze regulated by patriarchal social systems which seek to reduce her to an object and commodity. Through her representations, Ferdousi asserts the importance of studying not only the experiences of a *birangona* but also the lives of thousands of women who lived through the Bangladesh Liberation War of 1971. The study is one of critical importance to our understanding of the 1971 conflict, as it consistently re-emphasises the signs of women's agency in struggle, resistance and coercion.

Notes

1. There has been a selective representation of women's participation in the Bangladesh Liberation War. Women took active part in the armed struggle, took care of home and hearth and provided care for the sick and wounded. However, in the media representation throughout 1971–72, the focus remained only on those women who were sexually violated. It was only in the 1990s that researchers and activists like Nayanika Mukherjee, Yasmin Saikia, Bina D'Costa and Sultana Kamal brought out a more inclusive data set of women who had been a part of the war in various capacities.
2. For arguments regarding the problems of the historiography of Bangladesh see Ranabir Samaddar's 'Nation Building: Interpretations of Bangladesh War' in *Crossing Boundaries*.
3. Personal interview with Tariq Ali, trustee of the Liberation War Museum on 19 May 2017 in Dhaka.
4. Tk or Taka is the currency of the People's Republic of Bangladesh.

Works Cited

Ali, Tariq. Personal interview. 19 May 2017.

Debnath, A. "The Bangladesh Genocide: The Plight of Women". *Plight and Fate of Women During and Following Genocide*, edited by Samuel Totten, Transaction, 2009, pp. 47–66.

D'Costa, Bina. *Nation building, Gender and War Crimes in South Asia*. Routledge, 2011.

Hassan, M.A. *War and Women*. Dhaka, Tamralipi, 2010.

Hastrup, Kristen. *Other Histories*. Routledge, 1992.

Hirsch, Marianne, and Valerie Smith. "Feminism and Cultural Memory: An Introduction". *Signs*, vol. 28, no. 1, 2002, pp. 1–19.

Hossain, Hamida, and Amena Hossain, editors. *Of the Nation Born: The Bangladesh Papers*. Zubaan, 2016.

Islam, Kajalie Shehreen. "Breaking Down the *Birangona*: Examining the (Divided) Media Discourse on the War Heroines of Bangladesh's Independence Movement". *International Journal of Communication*, vol. 6, 2012, pp. 2131–48.

Mookherjee, Nayanika. "Lessons of Consent and Critique from Ferdousi Priyabashini". *Daily Star*, 15 March 2018, https://www.thedailystar.net/opinion/tribute/lessons-consent-and-critique-ferdousi-priyabhashini-1548142. Accessed 29 Dec 2020.

———. *The Spectral Wound: Sexual Violence, Public Memories, and the Bangladesh War of 1971*, Duke UP, 2015.

Rashid, Azra. *Naristhan/Ladyland: Gender, Nationalism and Genocide in Bangladesh. A Research-Creation Project*. 2016. Concordia University, PhD dissertation.

Saikia, Yasmin. *Women, War and the Making of Bangladesh: Remembering 1971*. Duke UP, 2011.

———. "Ayub Khan and Modern Islam: Transforming Citizens and the Nation in Pakistan". *South Asia: Journal of South Asian Studies,* vol. 37, no. 2, 2014, pp. 292–305.

Rahman, Hasan Hafizur, editor. *Bangladesher swadhinata juddho dalilpatra*. 15 vols. Dhaka, People's Republic of Bangladesh Information Ministry, 1982–1985.

Rossington, Michael, and Anne Whitehead, editors. *Theories of Memory: A Reader*. Johns Hopkins UP, 2007.

Samaddar, Ranabir. "Nation Building: Interpretations of the Bangladesh War". *Crossing Boundaries*, edited by Geeti Sen. Orient Longman, 1997, pp. 219–27.

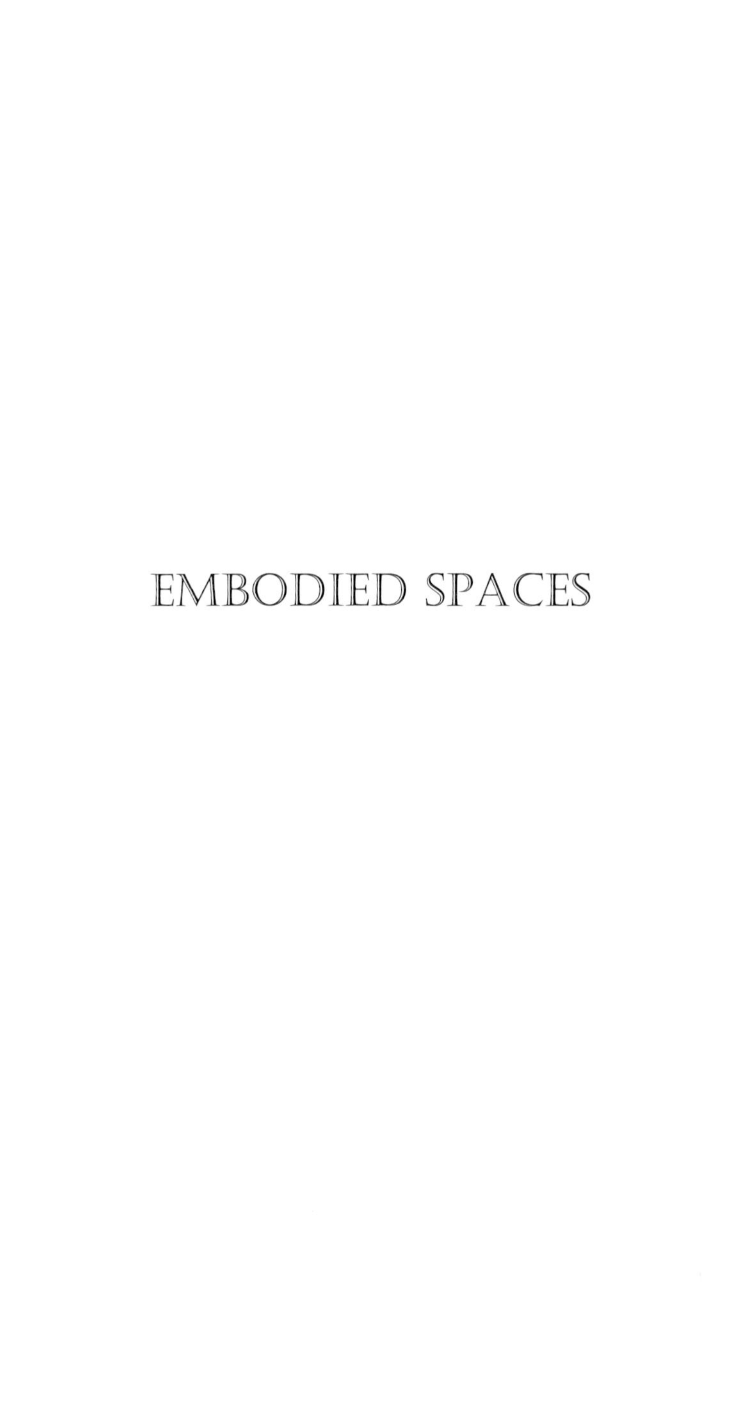
EMBODIED SPACES

ELEVEN

BECOMING THYLACINE

EMBODIMENT, ECOLOGY, AND MATERIALISM IN JULIA LEIGH'S *THE HUNTER*

Mark Byron

The current age of ecological crisis presents a fundamental challenge to the human imagination: how might a novelist capture a sense of this crisis and convey it to a reader, when the scale of the crisis exceeds the limits of vision? Can the singular conscience apprehend global crises by mediating the embodied self within an ecology of things, of species and individuals? How does one write of—or conceive of—processes of extinctions that accommodates both the species and its individual members? This essay surveys the writing of extinction and its capacity to stimulate reflection on systems and events too massive to allow direct contemplation: climate change, genocide, global mass extinctions. It takes as its focus Julia Leigh's novel *The Hunter* (1999) in which the last member of a species thought extinct for generations—the thylacine or Tasmanian Tiger—is discovered and executed for its reproductive and genetic material. Leigh's novel navigates planetary forces of species extinction in its attention to the singular and final member of the species, casting the reader's attention to the shadows of other extinction events of foundational significance to human identity: foremost among them the genocide of the Indigenous population of Tasmania by invading settler communities. By mediating barely conceivable forces of historical cruelty through the body of the hunter and the last thylacine, the novel brings into sharp relief interhuman and interspecies relations, providing a basis for contemplating and reimagining the modern human relation to ecologies more widely.

ECOLOGICAL EMBODIMENTS AND THE HUMAN-ANIMAL RELATION

As the global community is slowly waking up to the consequences of a century of oil and plastics, the pollution of the oceans and the atmosphere, and the inevitable rise in global temperatures, authors have developed strategies to absorb new theoretical work in their consideration of ways in which human subjects interact with and influence their societies, ecologies, and the world of objects. Perhaps the biggest lesson to have arisen in recent theoretical work across a range of disciplines—philosophy, ecology, psychology, sociology, and so on—and to have been absorbed in a raft of recent fiction writing, is the inherent and complex interconnectivity that structures our world and our movements within it. The COVID-19 crisis most vividly illustrates this structuring feature of our world, where global travel, government biosecurity regulations, epidemiology, preventative medicine, vaccination production and distribution, product supply lines, stock exchanges and rational and irrational behaviour towards our fellow humans all impact lived experience in profound ways. This extends to basic concepts of ecology too, in cases where governments seek to outlaw live animal trade in an attempt to counter cross-species transmission of pathogens, an unfortunately common factor in many of the world's recent epidemic outbreaks. The human–animal nexus is both a bridge and a divide—the key is to identify which particular relation is bridge and which is divide, and to act as respectful witnesses to the natural world and the limits of human intervention within it.

This focus on embodiment and subjective states has shaped what has come to be known as the ecological turn in literary studies, stemming from the ecocritical revolution of the 1960s engendered by such pioneering texts as Rachel Carson's *Silent Spring* (1962). Ecological themes have played a prominent role in Australian literature from its earliest incarnations—for millennia Indigenous stories and narratives have performed the grounded identity of people and country, and from the time of European invasion and colonisation two centuries ago literary texts demonstrate an almost compulsive preoccupation with land and ecology. The persistence of ecological crisis in literary expression bears a renewed topicality in the recent national trauma of the 2019–20 bushfires, in which dozens of people and approximately three billion animals were killed and at least 19 million hectares of forest were destroyed across every state in the nation. The somatic reaction to the fires described a spectrum, from the reactive immediacy of firefighting to reactions to the choking smoke enveloping cities with some of the worst global pollution measurements, as well as a complex array of post-traumatic stressors compounded in the population by the onset of the COVID-19 pandemic only weeks later. This compound event—two major sequential and overlapping disasters—only serves to accentuate the customary backdrop of natural disaster in Australian literature, and particularly the novel, throughout its history: it rarely functions as a consolatory force and in its often-bleak pessimism modulates fortitude and despair. How are bodies, affective states and the natural world scrutinised and arranged in such narratives, and what are the implications for understanding the capacity of narrative fiction to address the big questions facing the globe today?

These questions demand new ways of thinking about concepts of identity and affect mediated through ecology. If this sounds familiar, it is because we have been here before: the so-called *pathetic fallacy* first identified by John Ruskin and considered to be a signature feature of British Romanticism. This fallacy took the natural world as a mirror for human subjectivity, mapping human emotional states onto such phenomena as weather events: the entire *sturm und drang* tradition imported from German Romanticism and rehearsed in Book 1 of William Wordsworth's *Prelude* is one of innumerable examples; another is the recuperative affect tied to sunshine, in S.T. Coleridge's poem 'This Lime-Tree Bower My Prison', for example. What is new in the ecological turn is the way narratives are conducive to readings that esteem non-human agents, and even objects as agents in their own right: what Jane Bennett calls the *deodand*, an intermediary between human subjectivity and the objects of the world (Bennett 9). Julia Leigh demonstrates a keen critical awareness of the pathetic fallacy, giving her protagonist numerous opportunities to catch himself falling for its charms.

The specifically Australian inflection of the human–object relation is a self-conscious critique of the nation's aesthetic ideology. Australia was invaded and colonised when British Romanticism was in full swing, shaping attitudes towards a new and exotic ecology that proved both an existential threat and a zone ripe for exploitation. This is evident everywhere in colonial poetry and painting, where British sensibilities are overlaid upon utterly incommensurate landscapes. The nation still labours under this ideology, it having become the central nervous system of the (non-Indigenous) human's place in the Antipodean world. Leigh is one of a large cohort of recent writers to critically evaluate this attitude to ecology, opening up new ways of thinking through the human–ecological relation.

This reconsideration of the human–ecological relation is marked by renewed attention to how affect shapes the actions and attitudes of characters, and how it imbues itself in the very textures of narration. The affective turn in literary studies arose in the early 2000s alongside renewed interest in phenomenology, especially the work of Maurice Merleau-Ponty (see *The Phenomenology of Perception*). Drawing on the pioneering work of Silvan Tomkins, such scholars as Lauren Berlant and Sianne Ngai explore how theories of affect developed from the foundational role of phenomenology and its influence on later generations of theorists, who developed ways of configuring affect freed from the particularities of individual sensibility and emotion (see Tomkins; Berlant; Ngai; and Gregg and Seigworth). Brian Massumi theorises affect as a range of states independent from individual interiority, existing in the relation between subject and world and translatable across different subjects (Massumi, "The Autonomy of Affect", and *Politics of Affect*). In affect's distinction from emotion, he follows the seventeenth-century philosopher Baruch Spinoza. In this sense, Massumi's work pivots upon scholarship in medieval and early modern studies regarding the passions in literary texts such as Susan James's ground-breaking work *Passion and Action: The Emotions in Seventeenth-Century Philosophy* (see also Boyde). The debt to phenomenology, especially Merleau-Ponty's notion of an 'openness to the world', is also very much in evidence. Affect theory offers a means to evaluate how narrative subjects are open to their world in varying ways—by relations with nonhuman actors as well as objects. This awareness may induce readers to engage with these narrative subjects in new ways, in turn prompting new relations between the reader and their world and particularly with regard to some of its more pressing concerns: ecological crisis, human subjectivity and the dignity afforded non-human agents.

Merleau-Ponty's phenomenology is based in a rejection of an assumption of a readymade world underwriting both the idealist and empiricist worldviews, and rather, turns to the constitutive role of perception and embodiment. The subject faces world 'as it is' rather than as rational construct. This procedure situates phenomenology as an inheritor of the early modern reanimation of perception, in which it also performs a critique of the post-Cartesian mastery of world via the intellect. The relation between human subject and world 'is an ambiguous one, between beings who are both embodied and limited and an enigmatic world of which we catch a glimpse (indeed which we haunt incessantly) but only ever from points of view that hide as much as they reveal, a world in which every object displays the human face it acquires in a human gaze' (*The World of Perception* 54).

Merleau-Ponty also discloses how mature science opens the rational subject to gaps in knowledge and in the field of experience, inviting inner fantasies, dreams, and private interactions (56). This produces a pathway to the inner lives of animals, in which Merleau-Ponty regards Indigenous knowledge practices as crucial modes of this human–animal nexus.

Yet the category of the human itself demands interrogation. Taking Aristotle's notion of nutritive life in *De Anima* as a basis for subsequent divisions into lower and higher forms—'vegetal and relational, organic and animal, animal and human'—Giorgio Agamben identifies the caesura that defines the human as well as binding it to the economy of animal life from which it is set apart (Agamben 14–15). Thus, *Homo sapiens*—taken as an imperative to 'know thyself'—only becomes itself by realising it is other: *Homo* stems from the order of apes, but its very act of recognition sets itself apart. If speech is the characteristic that Karl Linnaeus identified as quintessentially human, its historical evolution rather than its innate inherence—evident in the problematic category of the mute man–ape or *Homo alalus*—means that the talking primate stands outside of the taxonomy of both animal and human: 'The animal-man and the man-animal are the two sides of a single fracture, which cannot be mended from either side' (Agamben 6). The human subject's attempt to mediate the human–animal nexus operates in this context, where both intimacy with and separation from the animal defines the human as the species set apart. This will come to define the nature and success of the quest in *The Hunter*.

BECOMING ANIMAL: THE LOGIC OF EXTERMINATION IN JULIA LEIGH'S *THE HUNTER*

The Hunter traces out a deceptively straightforward narrative trajectory. M, a university research scientist attached to an unnamed laboratory, is sent on a mission to capture the DNA of a thylacine or Tasmanian Tiger recently sighted in the island's Central Plateau. Thought to have been hunted to extinction when the last specimen died in captivity in 1936, recent unverified sightings of the thylacine have raised it to mythical status in Australia's faunal pantheon. The sighter, Jarrah Armstrong, has recently disappeared, presumed dead. M lodges with the deceased man's wife—who suffers deep depression and barely registers her own sense of subjectivity—and befriends the two children, Sass and Bike, who run wild in and beyond the house but are friendly and warm to

M on his stays. He takes 10-day hikes into the Central Plateau where he sets traps and submerges himself into the local ecology to wait on a potential capture. M is called away to Sydney unexpectedly, and during his absence Sass is badly injured in an accident in which her bedding catches fire. On his return several weeks later, M discovers that Lucy Armstrong has broken down completely and has been institutionalised, and Bike has been fostered out to relatives. He returns to the scene of his hunt to find that the National Parks and Wildlife Service has sent scouts to the same area to follow up on a recently reported sighting, but in tracking the scouts he learns they are not interested in their task and spend their days idly. M sights a female thylacine from the cave he inhabits, follows and kills it, taking blood samples and removing its reproductive organs, leaving the carcass for Tasmanian Devils—the largest extant carnivorous marsupials, *Sarcophilus harrisi* ('flesh-loving')—to consume. He leaves the area with his specimens stored in high-tech nitrogen packs, bypassing the two unsuspecting scouts on his way out of the forest.

The novel is written in a plain, declarative style when narrating the thoughts and actions of M. He introduces himself as Martin David to the Armstrong's but the narrator consistently refers to him as M. The narration occasionally turns to florid prose channelling M's description of the landscape or the atmospherics of different times of day, providing a kind of belatedly Romantic residue upon the landscape. The narrative point of view is in the third-person but swings between subjective and objective voices, almost always centred in M's perceptions and thoughts. One of the few times a striking subjective voice emerges is when M approaches the house after his prolonged absence and admits to feeling nervous. This shift in narrative voice heightens the impact when M finds the house abandoned and learns of the tragedy that befell its inhabitants. The final hunt sequence is reported in an objective voice, with M's thought processes and strategies taking on a forensic mood at the expense of any consideration of the ethical impact of his actions.

These narrative techniques undergird the deep moral ambivalence in M's activities as well as the implied moral economy submerged almost as deeply as the thylacine in the Tasmanian wilderness. The disappearance and presumed death of Jarrah Armstrong follows his reported sighting of a thylacine, and the accident that befalls Sass Armstrong raises a suspicion that their bad luck may be due to some unstated malign influence. They live on the outskirts of a small rural community notable for its antipathy toward environmentalists and hostility to outsiders. It is a place of last things, establishing the themes of extinction and extermination, a bluestone house in the shadow of the Central Plateau escarpment: 'the last house at the end of the last road' (Leigh 5). Father and daughter share non-native arboreal names—Jarrah (*Eucalyptus marginata*) is native only to southwestern Australia and no species of sassafras is native to Australia at all—which only signals more clearly that this suspicion may have some grounding. M's assignment is left undefined beyond that of the kill and sample collection as its goal: the reasons why a laboratory would want to acquire the genetic material of a marsupial long presumed extinct, rather than a live capture, are never revealed, and themes of genetic engineering are left unstated but ominously present. M's affinities with the land grow quickly through the narrative—he begins by calling a stand of blackwood trees on the escarpment track leading up to the Plateau 'the middle of a crowded nowhere' (13)—and his character invites certain kinds of sympathy as he navigates the extreme terrain of the Central Plateau. Yet his clinical execution and dismemberment of the thylacine tears away these readerly affinities with shocking directness. The relation of human subject to natural environment remains the central theme of the novel, where extended descriptions of M's 'becoming animal' are counterbalanced by the indifference and apathy of National Parks and Wildlife Service employees and the aggressive attitudes of the town's loggers and miners.

The majesty of nature is given priority in the narrative only to have its crown jewel murdered before the eyes of the reader in the novel's final pages. Any sense of the dignity of wilderness, the forest and its fauna, is remorselessly torn away ultimately demonstrating the deep hypocrisy embedded in an inherited Romantic ideology and its reverence for the sublime. *The Hunter* invites symbolic and perhaps even allegorical interpretation in the way it presents a deep human ambivalence to ecology. In one sense this is the history of colonial settlement of Australia, whereby British agricultural practices were introduced to often catastrophic effect and the endemic value of the continent's varied ecosystems brushed aside in favour of cotton, wheat, beef, and sheep farming. But there is another related discourse at work in this narrative, perhaps bearing even more insidious implications in its deep submergence. The characters in the novel are presented more or less stereotypically as Tasmania's modern population. None are Indigenous, and none show even the slightest signs of interest in anything related to Tasmania's unique Indigenous history—even the commune of hippies who establish a camp on the Armstrong property during the course of M's quest. The thylacine is a zoological cognate of the full-blooded Indigenous Tasmanian, hunted to extinction with even greater efficiency in the nineteenth century. M's stalking and pursuit of the thylacine through the Tasmanian high country echoes the infamous Black Line formed by two thousand settlers, convicts, and soldiers in October 1830, in which a front moved south and east across Tasmania to corral existing tribes into the Tasman Peninsula, designated as an Aboriginal Reserve by Governor George Arthur (after whom the Port Arthur Penal Settlement was named). The Indigenous Tasmanian population was decimated by massacre and disease from a pre-colonised population of perhaps 15000 people to the death of the last full-blooded Indigenous Tasmanian, Truganini, in 1876.[1]

Romantic ideology thus has a dark absence at its centre, the logic of total domination over an exotic ecology. This domination functions as the converse of the *horror vacui* and keeps it in check—the overwhelming experience of massive topographical features such as mountains, valleys, and oceans, often raised to metaphysical strata in the writing of Edmund Burke, William Wordsworth, Herman Melville, Percy Shelley, among many others. M is struck by the sight of vast valley floors, and from one high vantage point speculates on what he sees as 'what? Primordial civic planning' (30). M's reaction hints at the dark absence of human inhabitation and provides a direct thread between the historical fact of Indigenous dwelling and the thylacine. In his only direct reference to the land's first people, M notes a subtle human intervention in the land, leading him to a direct association with human extermination, legally sanctioned but which haunts the land. His attentive eye catches a ring of blackened stones, imagining it as a construction of 'the local Aboriginal people' before the final genocidal effort of the colonisers. M knows his history, recalling that the government attempted to create an Indigenous 'sanctuary' on De Witt Island, only to have this feeble consolation fail the same year as the last thylacine died in Hobart's Beaumaris Zoo in 1936. M also makes the connection between marsupial and human, recalling that De Witt has been proposed earlier as a sanctuary for the last remaining thylacines. This meditation is punctuated by the physical gesture of picking up a smooth stone and returning it to the earth (57). M's reaction here suggests a receptivity to larger forces of history and ecology, opening up a discourse that drives the narrative to its inhuman conclusion. The thylacine's extermination is foretold in the history of colonial displacement and massacre.[2]

M's quest for the last member of a presumed-extinct species lends the narrative an almost metaphysical tone, with echoes of Herman Melville's *Moby Dick* suffusing the hunt. M thinks of extensive searches for the thylacine undertaken by the World Wildlife Fund as well as 'private buccaneers and pipe-dreamers' (37) and rehearses the verdict of zoologists that the animal's extinction is assured by virtue of habitat fragmentation, competition with introduced apex predators, disease and hunting (37). M takes on the challenge as a latter-day Ahab, honing his 'all-seeing eye' and exercising 'his own godliness' (37). The direct reference to Ralph Waldo Emerson's essay 'Nature' is conspicuous here: M's rapacity is antithetical to the Emersonian respect for the natural world and the desire to view it transcendentally, and he betrays himself as an agent of instrumental reason in hunting down the last member of a species presumed extinct for the purposes of organ harvesting and genetic lab work. Rather, M occupies the material stratum of blood and bone where the cruel shock of his killing the mythic animal and extracting its genetic material recalls such speculative narratives as *Jurassic Park*. As Russell Smith has deftly elucidated, M's deep betrayal of nature and its human subjects in the administration of science devoid of bioethical considerations also calls up the spectre of *Frankenstein* (Smith 106–26). But in the final assessment, the ecological outrage committed by M in his obedience to his employers in a murky biotech industry is the epitaph to the human outrage committed a century earlier upon Tasmania's first peoples. The land, its fauna and its people are inseparable and the violence of colonisation reaching its logical conclusion in the killing of an unlikely survivor bears a symmetry impossible to ignore, much as history has attempted to do so.

How do the novel's ecological themes intersect with its considerations of affect and identity? M's affective economy consists of metaphysical speculation and literary references embedded in his response to the land. At various times during his quest, he fancies a kind of fusion with the ecology around him. When he observes 'the stars tak[ing] their infinite curtain-call', M feels his body dissolve and merge with the land: 'he is nowhere, everywhere' (82). This Emersonian expansiveness brushes up against the Christian theological doctrine that 'divinity is located with its centre everywhere and its circumference nowhere.'[3] M first feels his body grow light in its physical intimacy with the land. He then turns to the physiological minutiae of respiration, 'sens[ing] the air cool as it flows over the moisture in his nostrils' (82), returning him to an embodied sensorium required to fulfil its quest in hunting down an adversary. The plateau signifies the Romantic ideology of nature, becoming a place 'calm and pure, with space enough for a man to think' (139). M deploys these stereotypes to contemplate the geological timescales of volcanoes, earthquakes, and meteors, establishing an affective relation whereby the human lifespan is found wanting in its brevity. He concludes his meditation by asserting 'if a man's life were an island it would be uninhabitable' (139), conflating John Donne's famous declaration that 'No man is an island' with the scene of Shakespeare's *Tempest*.

M's attachments to the human sphere is evident in a range of literary and Biblical references, such as when he laments his situation climbing up the escarpment: 'I have been forsaken ... I did not ask for much, did I, no, other men have asked for more, and still I am denied' (141). This conflation of Job, Jeremiah, Christ's final words at the crucifixion and Peter's denial of Christ are put to the service of abject self-pity. However, upon reaching the plateau, a phase shift occurs beneath 'a pale moon', where M turns from metaphysics to the base materiality of disguise, making a paste of wallaby scat to cover his human scent (142). Stretching out his arms and imagining himself airborne, M leaves behind the Romantic ideology of nature and assumes the phase of 'becoming animal'. This is his weapon against his adversary: to occupy her world and assume its affects and, thus, to lurk undetected until the time to strike is upon him.

The process by which M becomes animal begins with familiar demonstrations of bushcraft: M knows to cover himself in a paste of animal dung to disguise his scent (30); he marks trees and builds cairns to guide him in and out of the forest; and he sets traps and snares with precision and subtlety. His acute sensitivity to the region's ecology and geomorphology is evident in his 'impersonating' a bee to trigger a flower into release its pollen (28); his identification of various animal scats such as wombat cubes, wallaby pellets, and 'devil twists full of hair and bone' (28); his identification of pawprints in the muddy earth (92); and his deliberations on which pad to follow by falling on his hands and knees and sniffing the air (113). Yet this 'becoming animal' never quite casts off his predatory human identity: M remains acutely aware of his location and of the way animals track through the forest (117). His actions lead M to discover human remains hidden in the undergrowth, presumably those of Jarrah Armstrong. This pulls him out of 'becoming animal' and induces a quintessential human response of erecting a makeshift grave marker: 'Here lies J.A. May he rest in peace' (115).

M also observes the steep cliffs of the escarpment as dolerite deposits (99). These deposits are the world's largest, having been produced by massive vulcanism during the breakup of the Gondwana supercontinent in the Jurassic period roughly 180 million years ago. Dolerite also produces bluestone, the material from which the Armstrong house is constructed (and is, incidentally, the building material of Stonehenge, that icon of British deep time). M 'performs his favourite trick' on the dolerite platform, when 'he changes shape, swallows the beast' (91), figuratively growing thick fur and a tail. M's attempts to merge with the ecology produces an affective chain of dream and memory, where a potent emblem of the thylacine is located in the ghosts and the dreaming of long ancestral animal memory seeping into M's consciousness. He asks himself 'Do tigers dream?'—and do they dream of a mate's scent, or as the quarry of an unseen predator (45)? M engages the thylacine in imaginary conversation—'romancing his prey' (90)—as though able to access the hereditary memory of the entire species, epitomised in its last specimen. This approach is mediated by folk wisdom of the thylacine's threat to pastoralism, where roaming 'yellow-eyed packs' conflate perceived threat with fairy tale: 'You'd drag babies out of their frilly cots and wolf down frilly girls whole ...' (48). When M discovers a quoll in one of his snares his inhuman response—watching the animal's death throes 'as if his earlier dreams had turned to wax in his ears' (54)—calls up the image of Odysseus lashed to the mast of his ship as it passes the Sirens, only to erode any sense of literary elevation when M flings the dead body into the scrub with a complete disregard for its dignity as a once-living being.

Yet M's 'becoming animal,' for all its material success, falls short of genuine affective sympathy and instead reveals itself as merely a highly effective process of mimicry. M's technical engagement with the material facticity of his world means that individuation wins out in his execution of the thylacine. This marks his breach from an affective relation with his ecology, deferring instead to the biotech narrative lurking at the edges of the novel. M muses on the skill or luck possessed by the thylacine, a hardy remnant of her species, wavering between an image of her mangy species degeneration and a nobility that makes her a worthy adversary. M too, is revealed to be the last of his own line—the narrative provides a brief glimpse at his romantic history with mention of a girlfriend and a terminated pregnancy in his youth (69). His discovery of a print soon engages his taxonomic expertise to identify it as belonging to the thylacine due to the size of its plantar pad, the number of toes, and the gap between claw, digit and main pad comprising 'the deciding factor' (92). M draws from the reservoir of human *techne* as he waits for his quarry, 'carving designs into bones' with the point of his knife (160), as well as gathering flowers and feathers to decorate his lair. He begins to invent new words for natural phenomena but soon gives it up in favour of monastic silence (160), becoming the paradoxical *Homo alalus* whose silence marks him off from the animal realm but also inscribes an inner division that defines his humanity. M's final application of technique and skill occurs when he executes and eviscerates the thylacine removing her reproductive organs, but he is unable to shake the bare fact of the gulf dividing the living from the dead: 'he observes her body as he would the body of a friend laid out in the morgue', reflecting on how 'her stillness is obscene' (164). The obscenity is M's, though, as he desecrates the long human history of marking death, where the gravesite produces a rich symbolic economy from the horrifying absence of human passing to be passed down to future generations in a history of affect—the lifeblood of collective and individual identity.

M engineers the anthropogenic extinction narrative in *The Hunter* in his ability to mute his species distinction and to mimic the process of becoming animal. For all his musings on the dreamlives of tigers and his crypto-Romantic responses to the sublime wilderness of the Tasmanian Central Plateau, M's execution of the last thylacine also kills off any consolatory affective relation to the novel's ecology, installing what Agamben calls the 'zone of indifference' at the centre of the modern 'anthropomorphic machine' (Agamben 37). M takes a temporary vow of silence—likening himself to a monastic observing the *magnum silentium*—in a gesture that positions him as a parasite upon the ecology in which he is embedded. This silence also marks out an inner schism in which the pursuit of his quarry means divesting himself of a fully self-reflexive humanity. M becomes *Homo alalus*, the silent man-ape whose ambiguous status calls into question the category of the human per se. His reflexive gesture in marking the resting place of Jarrah Anderson—where the Greek word σεμα (*sema*) means both 'grave' and 'meaning'—is an empty gesture in light of his disrespect for nonhuman animals at the scenes of their deaths. As M's killing of the individual thylacine marks the extinction of the species, so does his simulated respect for Jarrah Anderson cast into relief the blithe disregard for Tasmania's Indigenous inhabitants. By scaling its narrative at the level of individual agents—both human and animal—*The Hunter* illustrates how simulated and empty affect provides a stark resource with which to face the hyperobjects of mass extinction, climate change, and genocide.

Notes

1. For a study of British government policy and ideology during the colonial era in Tasmania, see Lawson's *The Last Man*. Cassandra Pybus has produced a long-overdue biography of 'the last Tasmanian Aborigine' in Truganini. The foundational account of Indigenous resistance to European colonisation across the Australian continent and Tasmania is Henry Reynolds' *The Other Side of the Frontier*.
2. The only other significant reference to Indigenous habitation of the land is made by the child Bike, who tells M, on one of his visits, the story of how the thylacine gained its stripes. Bike recounts how the spirit *palanna* was saved from an attacking kangaroo by a thylacine joey (the term for juvenile marsupials). On returning home, *palanna* mixed some blood from the fight with the ash from a fire and painted stripes on the joey's lower back to signify his bravery. This story closely follows that of the Nuenonne nation of Bruny Island off the Southeast coast of Tasmania, as told by Leigh Maynard in 2007 (see Maynard 2004).
3. The earliest textual record of this aphorism is the twelfth-century Laon manuscript of the *Liber XXIV philosophorum* (*Book of the 24 philosophers*), a text variously attributed to Hermes Trismegistus, Aristotle, and the fourth-century grammarian Marius Victorinus. It appears subsequently in the writings of Alan of Lille; in the *Romance of the Rose*; Dante's *Commedia* (*Paradiso* XXXIII); and Rabelais' *Gargantua and Pantagruel*; and is quoted by Marsilio Ficino, Nicholas of Cusa, Giordano Bruno, Blaise Pascal, Voltaire, and Friedrich Nietzsche, among others.

Works Cited

Agamben, Giorgio. *The Open: Man and Animal*. Translated by Kevin Attell, Stanford UP, 2004.

Bennett, Jane. *Vibrant Matter: A Political Ecology of Things*, Duke UP, 2010.

Berlant, Lauren. *Cruel Optimism*, Duke UP, 2011.

Boyde, Patrick. *Perception and Passion in Dante's 'Comedy'*, Cambridge UP, 1993.

Carson, Rachel. *Silent Spring*, Houghton Mifflin, 1962.

Gregg, Melissa, and Gregory Seigworth, editors. *The Affect Theory Reader*, Duke UP, 2010.

James, Susan. *Passion and Action: The Emotions in Seventeenth-Century Philosophy*, Clarendon, 1997.

Lawson, Tom. *The Last Man: A British Genocide in Tasmania*, Bloomsbury, 2014.

Leigh, Julia. *The Hunter*, Penguin, 1999.

Massumi, Brian. "The Autonomy of Affect". *Cultural Critique*, vol. 31, 1995, pp. 83–109.

———. *Politics of Affect*, Polity, 2015.

Maynard, Leigh. *How the Tasmanian Tiger Got Its Stripes*, illustrated by Aboriginal Nations Animation Studio, Scholastic, 2004.

Merleau-Ponty, Maurice. *The Phenomenology of Perception*. Translated by Colin Smith, Routledge, 2002.

———. *The World of Perception*. Translated by Oliver Davis, foreword by Stéphanie Ménasé, introduction by Thomas Baldwin, Routledge 2004.

Ngai, Sianne. *Ugly Feelings*, Harvard UP, 2007.

Pybus, Cassandra. *Truganini: Journey Through the Apocalypse*, Allen & Unwin, 2020.

Reynolds, Henry. *The Other Side of the Frontier: Aboriginal Resistance to the European Invasion of Australia*, History Department, James Cook University, 1981.

Ruskin, John. "Of the Pathetic Fallacy". *Modern Painters*, vol. 3, part 4, Smith, Elder & Company, 1856.

Smith, Russell. "Global Modernity, Anthropogenic Extinction, and the Future of Sexual Difference: From Mary Shelley's Frankenstein to Julia Leigh's *The Hunter*". *Affirmations: of the Modern*, vol. 4, no. 1, 2016, pp. 106–126.

Tomkins, Silvan. *Affect Imagery Consciousness*. Springer, 1962; 1963; 1991; 1992. 4 vols.

TWELVE

WALKING ON A TIGHTROPE

CIRCUS AND ITS DOUBLE IN COLONIAL BENGAL

Prodosh Bhattacharya

The turn towards the body in the dialectic of mind and body presents a broad field of investigation for performance studies. The anti-theatrical turn in theories of performance liberates the body from its art-bound anatomy to a structure of feeling. Bodies in this context might become a lens for or a methodology of cognition. In this trajectory of embodying knowledge, it is crucial to ask what constitutes the event of circus. The emergence of 'print capitalism' (Anderson 39) as a crucial factor in the nineteenth and early-twentieth century discursive formations of colonial identity in India, marks a movement towards the textualisation of the visual—inscriptions of ways of seeing. If we delve deeper into this argument, it invokes two critical positions: (a) the form of circus and its staging of the colony; (b) colonial discourse formation through the textual circulation of the empire and the extent to which it mediated the performative dimensions of circus. The former position is elucidated by Paul Bouissac:

> Circus performances, though, are essentially visual and acoustic events. "Writing the circus" cannot merely consist of producing reliable verbal narratives of circus acts. Even if these translations into a written language can achieve some degree of completeness when they are supported by exhaustive analyzes of the components and dynamic of the multimodal processes of these acts, the visual information will always be missing. (*Circus as Multimodal Discourse* 29)

Within the field of circus studies, situating circus in colonial Bengal as an embodied space goes beyond considerations of circus as an anticolonial, nationalistic engagement of the body. This chapter opens up ways of looking at the form of the travelling circus beyond its structural and organisational apparatuses and seeks to engage with its visceral dimensions in a capillary critique of the cultural techniques of improved corporeality as a discourse. In this regard, the ideologues of colonial modernity and its framing of bodies is fraught with tensions that are marked in an affective reading of circus in colonial Bengal as a space of intersection between the textual and the bodily aspects of its physical culture.

While the nature of corporeality in a performative context imbibes phenomenological attributes to a considerable extent, it is significant to consider how bodies engender performance spaces along historical, gendered and biopolitical axes of power. Circus in colonial Bengal can be conceptualised as an area that aptly elucidates the nodal points of corporeality and the discourses around it, through a deep and affective reading of the textual and visceral dimensions of circus as an embodied space. This could give rise to fresh debates around existing circus historiography.

NARRATIVISING BODIES UNDER THE BIG TOP

The performative dimension of circus as an event is marked by physicality in both its personal and semiotic spheres. The semiotic bodies of the circus thus present themselves as addenda or extensions of the material bodies in the context of the spectacle of performance—'the tactility of vision is an addendum to the visible space' (Merleau-Ponty 134). The various spectral movements of the acrobats on triple bars, horizontal bars and trapeze as well as the freak shows, all add to the kinesthetics of circus as an event and an experience of the liminal staging of itself as an embodied experience.

While circus's performative visuality evades strict codes of theoretical and conceptual categories—phenomenology, neurocognitive science, semiotics, of how the idea of the moving body may be conceived within the framework of performance theory—writings on circus and its associated physical cultures have been in increased circulation since the 1850s. These moving and malleable bodies are thus stretched and contorted, not just in the arena, but also beyond it into text through memoirs; travel books; newspaper reports; quasi-historical accounts; biographies; narrative accounts of circus enthusiasts; and fiction. The mode of circulation of physical culture, not only in practice but also in the imaginative and fictional currency of these literary sources, becomes a necessary engagement in understanding how the body intersected with popular discourse. The loss of the visual in the modality of liveness, not as a phenomenal reality but as an ontological state, becomes necessary to engage with while thinking about circus through/in the written form.

A dominant narrativisation of body practices in nineteenth and early-twentieth century Bengal aimed at the deification of certain forms into a 'nationalist' strain—a form that was predominantly a middle-class Hindu one. The trope of the brave, masculine, heroic male body stood at fictional loggerheads with the popular notion amidst Europeans of Bengalis as a lazy and effeminate race. This strategic narrativisation forms one of the intersections between circus in text and performance, marking the metonymic abstractions that cultural forms take. Abaninda Krishna Basu presents a specimen of the European view of the Bengali character by referring to an entry in Mr. Haber's (a British colonial officer) journal: 'I have understood from many quarters that the Bengalees are regarded as the greatest cowards in India.' (qtd. in *Bangalir sarkas* 9). This also finds classic expression in the following observation by Macaulay:

> The physical organization of the Bengali is feeble even to effeminacy. He lives in a constant vapour bath. His pursuits are sedentary, his limbs delicate, his movements languid. During many ages he has been trampled upon by men of bolder and more hardy breeds. Courage, independence, veracity, are qualities to which his constitution and his situation are equally unfavourable. (25–26)

In this regard it is of historical and political significance to note as a counterpoint to the alleged enfeeblement of the Bengali race as an intrinsic trait, that in 1700, Calcutta had 1200 English men and women, 460 of whom died in one hot summer. This was largely due to a policy of pauperisation monitored by a court of British directors where poor, diseased and vagrant bodies were isolated, marking a systemic abjection in colonial politics. The management of bodies as resources of utility through labour was controlled through the Vagrancy Act introduced in 1870. The organisation of labour exchanges in workhouses in Bombay during the nineteenth century point to poor conditions of work and wages. It was through these workhouses that marginal Europeans were employed, but who could not compete with the rate of cheap labour that was available among the natives. This led to a growing discourse around vagrant Europeans—some of whom had been in India for a long time and those who had been born here—as corruptible, degenerate, feeble and morally unscrupulous due to the influence of the native races and climate.

In this period when there was a prevailing discourse of Bengalis as a weak race, mastery over the body as a means to attain a unity of self thus found narrative extension in works like *Anandamath* by Bankim Chandra Chattopadhyay and *Swapnalabdha Bharater itihas* (*The history of India as received in a dream*) by Bhudeb Mukhopadhyay. In the latter, the author addresses history and seeks to define the contours of nationhood in terms of a merging of Bengali, Hindu and Indian identities. The proliferation of upper caste and class narratives sought to bring about a dematerialised economy of feeling bodies within the framing narrative, thus echoing the complex interplay and dialectics between the spiritual and material domain of nationalism.

It is with Priyanath Basu's circus venture with a Bengali troupe—the Great Bengali Circus—that efforts were made to restore the spirit and body of Bengalis. Building itself on the foundations of the 'Hindu Mela', Nabagopal Mitra's National Circus and the proliferation of *akharas* (wrestling areas), the Great Bengal Circus presented a *mise-en-scéne* of colony as a form of mimicry, elaborated on by Homi Bhabha as 'almost the same but not quite' (Bhabha 126). This was possible through a system of drawing patronage from both British officers of the Company (as the East India Company was commonly referred to), and zamindars from and rajas of various districts and parts of the country.

However, a deconstructive mindfulness of this perceived idea of the Bengali resistance and resurgence through physical feats, is fraught with philosophical and cultural complexities. J. Rosselli makes a cogent observation: 'A long-standing ambiguity about the definition of Bengali society and the place of the elite within it coloured the myth of physical downfall and resurgence: excluded or hierarchically subordinate elements (Muslims and low castes) could neither be kept out of the story nor satisfyingly accommodated within it' (121).

Circus's own encapsulation of these identity formations should, in some way, make thinking about its form possible. While delicately strewn across the dimensions of work and play, its modern structure—from Philip Astley's formalisation of the modern circus in the eighteenth and nineteenth centuries—was coterminous with the industrialisation of time. Circus' own co-opting of industrial modernity itself hinges on its performative habitat that turns its nomadism into distinct place-making. It has a capacity for being a messianic space—a space of all spaces—a kind of temporalisation of space that takes the everydayness of time, locatable out of time. Its performative mode is a not a denial but a momentary stalling of history as a project of the state.

PERFORMING COLONIALITY: AFFECTIVE INTERFACES OF CIRCUS BODIES AND THEIR STORIES

It is imperative to ask what gave the travelling circuses in tents—both the ones that came from Europe to India, and Indian circuses—their distinct deified ideological constructions? How do two parallel events of circus mobility that are correlational, turn to causality of the said constructions? The question takes us further into the problematic intersections of the methodological conventions of history and literature, science and philosophy. Is it possible not to speak of one despite the other? A possible answer might lie in considering how the material economy of the objects and the props in circus have an active agency in marking identities, along interhuman, human–animal and human–object interfaces.

A dominant feature of Astley's circus was the equestrian show, featuring displays of horsemanship in the circular space as a marker of the supremacy of the empire. This aspect is grounded in the fact that these equestrian shows were mostly performed by soldiers and their families. The rehearsed showing of the rituals of controlling the horse consisted of graceful and acrobatic displays on the horse's back, also known as trick riding. In the systematised practices of European equestrian shows, a sartorial framing of these acts regulated the ideological through a disciplining of the body. It is interesting, however, if we ponder on how precariousness—as inducing an affect of fear and wonder—enacts the cultural symbolism of empire and its discourse of supremacy through the controlling of animals. Animality in circus ranged between conceptions of mastery, control and the disciplining of the wild, and a celebration of taming brutish wildness which stood in for an aggressive anticolonial spirit—a spirit that becomes the ideal bedrock of a national ethos. It is in this context that the notion of adventure wears an expansionist idiom in enterprises like the Great Wilson Circus, embodying the ethos of the empire across its colonies. However, indigenous circuses and their tales of physical feats enacted a mobility of escape through narratives enmeshed in the globalism of adventure. As Anirban Ghosh observes, 'The letters sent by Colonel Suresh to his uncle speak of a continuous denial of his life and community history in the native land; he spoke of "tackling white wrestlers and doing great feats never achieved by the lazy Brahmins in his village"' (11).

A critical consideration of the slippages that the human–animal interface might represent in the event of circus, presents the possibility of deflecting any simplistic binary interpretation of coloniality into cultural modes of the self and the other. While the sensory realm of circus brings to the viewer the specific sociology of the interpretative habitat that colonial Bengal presents, the bodily experience of wildness offers the affective deconstruction of circus as a cultural artefact. The momentary possibilities of transformation through the untamed excesses of the visceral experience of circus acts therefore spread into the wildness of the signifiers (circus lore, posters, fictional narratives) that, in their own ways, performed this apartness of the travelling circuses as a differential writing of identities in colonial modernity. These articulations themselves serve as expressions of the limits of modernity within which they are engendered. A performative engagement of the circus spectator's presence in the phenomenology of myriad bodies becomes a crucial function that marks circus's affective thrust as a category of embodied knowledge production. Peta Tait observes,

> Circus animal acts were promoted as "wild" and "dangerous", and spectators brought these emotional expectations to viewing them.... trained circus movement paradoxically contravened the way in which the body's appearance and action delineates the boundaries between species; as animal performers became anthropomorphised in the human-like actions of an act, human performers were accordingly "zoomorphized". (187–88)

In this context, the tales of bravado and adventures of Colonel Suresh Biswas become an apt example of a subject body, mobilised in a narrative frame of wonder and adventure that he enacts by being the animal tamer in transnational contexts. As an entry by an anonymous journalist, writing for *A Semana* newspaper puts it: 'A tiger, two lions and a man, kept in the same cage is something that will get the nervous system of the world's most peaceful citizen working. What fear, my God, what fear those two terrible lions and that no less terrible tiger create! And, meanwhile, the fearless Sureesh Beswash handles them like some domestic cats' (qtd. in Dutt 18).

The popular circuses of Europe that travelled to India had exotic shows where the lions and tigers had Indian names. The animalisation of the colonised (which the Empire makes civilised) is played out in these shows. This is epitomised in a circus poster featuring the act 'Garden of Eden' in 1889, from Calcutta. This becomes significant in Great Wilson's Circus and Chiarini's Circus which would present acts with names indigenous to the place they played in. Another such employment to this effect is observable in the strategic manoeuvre of rumours about tigers and leopards in the circus having escaped their cages and being astray in the town till Badalchand, the animal tamer,[1] successfully captured them. However, what such animal acts may successfully mask is the deeper affective violence that resides in the subliminal regions of civilising missions. As Anirban Ghosh observes in *The Tropic Trapeze*: 'By rendering the labour of the circus performer into a commodity the society sought to manipulate and represent the circus according to their wishes. The circus was a market of objects that worked according to the desired "demand" of the customer. This in turn erased the violence that went into providing for the "supply" to that demand' (6).

Circus as a performative principle brings into this commodified exchange an excess that cannot be changed into a value. While its structural components contribute to its assimilation of consumption into a visual narrative, its specific ontology makes its presence ephemeral to sustainable codification. Therefore, a possible model of the layered principle of circus performance and its association with value is to comprehend it as embedded in the nexus of ritual and as a commodified form of embodying a cultural space. A literary dimension to this idea is embodied in Shirsho Bandopadhyay's work *Shardulsundari*[2] which explores the possibilities and failings of love as values that evade the commodified form, and how Sushila Sundari is caught in its self-revelatory thrust. Looking at the narrative alongside Sushila's documented life in a male, upper-caste, Hindu, nationalist ethos places before us the possibilities of absent responses lost in the staged spectacle of Sushila with two Royal Bengal Tigers. The fictional account of Sushila's consumption is embodied in the episode of a tiger biting off flesh from Sushila's face, leaving her with permanent marks of voyeuristic undesirability. Her trajectory in history masks this violent possibility through discourses of brave Bengali women who are consumed by the voyeuristic gaze and the grand narrative of Mother India.

In the nationalist discourse, physical culture upheld the trained male body as an ideal repository of improved corporeality and development. In this regard, the engagement of the mobility of travelling circuses with narratives that are indicative of how circus as an institutional space sought to represent bodily skill, are implicit in the discourses of physical culture in colonial modernity. The representation and discussion of the body in nationalist discourse thus privileged a masculinist idiom of fearlessness and strength of character. To such an effect, it also framed the presence of female performers in physical culture as a production of male tutelage. It emphasised an already growing discourse of female iconography and representation as specifically visible in masculine and gentrified nationalist contexts. A mindful deconstruction of narratives and texts that embed the corporeal in circus settings, presents an important critique of how bodies possess resistive economies to means of repressive control by both the colonial state and its anticolonial counterpublic. A demonstrable example to that effect is an interesting account in the form of a travelogue by Professor Priyanath Bose in 'Circus e bhooter upodrob' ('The haunting at the circus') that presents, through the imaginative currency of language, the affective leakages of fear and loss of bodily awareness amongst professional circus performers who are believed to have mastered fear through shows of bodily prowess. We have Professor Bose recount,

> We all started to look at each other in panic, amongst the commotion, powerless to move. We tried to shout for help but could not do it. The next moment, we saw a chimera of five to seven terrible human forms, dancing about us; it made our teeth clatter with terror and we eventually fell unconscious.... when we regained consciousness, I found myself lying on a bed and some people gathered around me. Even till date, any reference to this incident makes our hearts palpitate and skip a beat. (*Professor Bose er apurbo bhromon brittanto* 194–96, translation mine)

ANUSHILAN (PRACTICE), MODERNITY AND THE AFFECTIVE GENEALOGIES OF BODILY REPERTOIRES

The epistemic frameworks of industrial modernity sought to invest the circus form with a rationalisation of skill as a project of the West, and appropriated indigenous and local practices into institutions of knowledge formation. The Bengali circus (in the early decades of the twentieth century) produced an alternate repertoire of the skilled body through an assembly of indigenous body practices and with the adaptive transformations of resilient bodies in a social environment of estrangement. As Brian Massumi opines,

> The physical dimension of the body corresponds to actions performed in conformity to the past, continuing along the same lines as it, following the

> same schema. Thus the physical is a principle of conformity to already-emerged form. What characterizes mentality is the capacity to go beyond that givenness to improvise new forms. Note that I said 'mentality', and not 'the mind'. Here, mentality is a mode of activity, and it functions not in opposition to the physical but with it and through it, by prolonging and renewing it. (179)

A central tenet of Massumi's argument, which holds critical currency in understanding the affective trajectories of colonial circus, is the possibility of flux and change in movements that are present prior to the positing of the thinking subject. In this regard a changing environment would thrust before the performera range of challenges to keep pace with. The body's capacity for resorting to an archaic yet ever present repository prior to independent and reflexive subjectivity therefore becomes the channel of adjusting to a bodily history.

Circus acts like trapeze, walking on tight rope, jugglery, *lathikhela* (stick play), are not subjective. They are a hollowing out of the subject based on contingencies of physical laws taken to its limit. To engage with a Freudian economy, each of these acts straddle life and death drives and in the ocular logic of the spectacle, they suspend the phenomenological to generate affective touch.

An important dimension in this regard is the composite form of the spectator–performer. To lay bare affective tactility in circus performance, one must be able to make a shift from the physiological body of the performer to mobile Duchampian extensions—of looking at the body as movement; of the trapeze artist in conjunction with the trapeze; of the walker along with the dimensionality of the tightrope; of the juggler with the performative economy of a three in the numerical possibility of two. As Antonin Artaud contends in *Theatre and its Double*, 'To be familiar with the points of localization in the body is thus to reforge the magical chain' (99). It is the story the body enacts on consciousness, as knowledge that acts upon one even prior to cognisance.

A critical tangent in this context is the conversation between how bodies may be historicised and the somatic registers of how circus embodiment works. Antonio Damasio theorises that bodies that are writing their shared experience and are reclaiming reflective principles of consciousness must negotiate the old Cartesian model of the mind–body problem through a neurobiology of consciousness. Embodied knowledge and its flow thus present an interesting counterpoint to epistemic thinking through subjectivity.

Pulin Behari Das's meticulous discourse on the repertoire of *lathials* (stick performers), traditions of sword fighting and knife work, marks an important moment in this bodily history of *anushilan* (practice). It was coterminous with the narrativisation of time at the cusp between the precolonial and colonial modernity. The emergence of print culture further enabled the establishment of the Dhaka Anushilan Samiti as the nodal point of discourses and practices of nationhood following the partition of Bengal in 1905. The embodiment of *anushilan* stands in contrast to what Michel Foucault characterised as docile bodies (127). Foucault's understanding of the body was in a biopolitical environment where it was individually subjected to subtle operations of control and power. That the discursive nature of extremist notions of nationalism had co-opted this bodily practice as political capital, underscores the anti-docile, performative and visceral tactility that informs it as a counterpoint to ways bodies are always already regulated in discourse.

It has been suggested that Pulin Das learnt the skills and techniques of the bodily forms in *lathikhela* from Abdul Rahim Popham, also known as Professor Murtaza, who was associated with the Great Indian Circus. It has been postulated that Murtaza had learnt the conventions of stick fighting, sword fighting and knife work from incarcerated *thogis* (thugs) while in a Pune jail (Bhattacharya and Ray 17). Pulin Das takes on the position of an ethnographer of physical forms as his writings inscribe the distinct observations he made while observing and training under Murtaza. The distinct taxonomical nature of his observations on the movements in a stick fight and on the codes of behaviour apt for revolutionaries marks an important moment in the history of *anushilan* as a performative disciplining of the body and mind to adapt to the accidental nature of experience. This is embraced in the saying 'Shorirer nam mohasoy, ja sohabe tai soy' ('the body is yours obediently, make it do whatever you like') which refers to the body's situatedness that marks its capacity to adapt to being acted upon.

In this regard, the performative iterability of the everyday points to the possibility and importance of political thinking through affect and reinvigorates deep-seated debates on the linear usefulness of affects that act on us and the inherent violence in the structuring consciousness as the meeting ground of and a validator of affects. Antonio Damasio, observes in *The Feeling of What Happens*,

> Consciousness begins when brains acquire the power, the simple power I must add, of telling a story without words, the story that there is life

> ticking away in an organism, and that the states of the living organism, within body bounds, are continuously being altered by encounters with objects or events in its environment, or, for that matter, by thoughts and by internal adjustments of the life process. Consciousness emerges when this primordial story—the story of an object causally changing the state of the body—can be told using the universal nonverbal vocabulary of body signals. (30)

EMBODIED SPACE: A HISTORIOGRAPHY OF COLONIAL CIRCUS AND SOMATIC TRACES OF MOTION

A possible way of finding the staging of colony is also to look at it as the double of circus that enacts life through affective transmissions, connoting gestures at the edge of death. The avenues afforded in exploring these primal states as staging specific colonial ideologues might then direct one to principles of methodological simultaneity of the archival and the ethnographic; of the historical and the performative; of the neurological and psychological, without collapsing their difference.

Arguing for circus's form as engaging multimodal transmissions may then open critical spaces to understanding it as a continual engagement with an evasive performativity. In its distinct colonial context, the circus in Bengal reifies a complex interplay of the narrative of the nation's claim on the body with a principle of nomadism that is endemic to itself. In colonial modernity and its operations on how it ordered and exacted the subject–body through discursive strategies, circus remains a signifying marginality that co-opted the circulation of militant nationalism as well as resisted the framework of a territorial mode of being through the transnational economy of its mobile performers.

An important line of inquiry lies in accounting for existing circus literatures and the veracity of their discourse in circus as an experience in itself. In more ways than one, both Pulin Behari Das's *Astracharcha* and Dutt's account of Suresh Biswas perform the embodied ephemerality of the circus on principles that are on the opposite ends of the spectrum of colonial embodiment. The taxonomic and coded text of *Astracharcha* coincides with the radical turn of practice to an extremist nationalism, and therefore co-opts the bodily repertoire of *lathikhela* into a mode of textual circulation that was part of the activities surrounding the Dhaka Anushilan Samiti (Bhattacharya and Ray 12–13). It is, however, interesting to note that the discourse surrounding *anushilan* marked a distance from the early nationalist writings on body practices as a mode of restoring the character of the Hindu, male, Bengali gentleman. Going beyond the politically deterministic readings of physical culture in early twentieth century Bengal, Pulin Behari's repertoire represents an assimilation of practices from the rural countryside and Japanese martial arts form in the textualised embodiment of the revolutionary political affects that the performance of subject–bodies hold in their precarious existence. The narrative temper of Suresh Biswas' life, on the other hand, upholds and sustains the literary–mythic quality of the transnational heroic subject of adventure. The always already fictionalised accounts of Suresh Biswas' life provide the affective rapture that circus as an experience generates.

It is thus crucial to obfuscate the neat boundaries of the phenomenal and the mimetic categories of circus and circus writing respectively, and to consider it as a methodology that activates thinking through affect as embodied in the colonial entanglements of modernity. The ambivalent and fragmentary subjectivity through the visceral and precarious bodies in colonial circus may thus find an explicative model in the ideological impasse between feeling and consciousness in the neurobiological formulations of homeostasis, according to which, while a vast array of feelings cuts across a body and its attendant consciousness the chemical variation in each cell is very narrow, as otherwise the body would die. This asymmetric functionality is the body's way of restoring stability amidst affective multiplicity. The embodied space of circus, through the primacy of the body as that which enacts and is inscribed upon, presents a methodology that may reconstitute the visceral as not just a totalisation but also a capillary mode of thinking. In such a capacity, colonial circus in Bengal posits a deep and slow reading of how the traces of body subjectivities perform and constitute the colony and its cultural relations.

Notes

1. Badalchand, also known as Bir Badal, was an animal tamer at Professor Bose's Great Bengal Circus in the late-nineteenth and early-twentieth century. He was, as several newspaper accounts and circus narratives suggest, a deft performer and mastered the art of taming lions and tigers, marking a bestial affinity that these ferocious beasts lend themselves to and turning the human–animal relationship into one of playful spectacle. For more details, see *Bangalir sarkas*, 209–10.

2. *Shardulsundari* is a novel published in 2016 and written by Shirsho Bandyopadhyay, set in the Great Bengal Circus of the late-nineteenth and early-twentieth century. It imaginatively constructs a plot around Sushila Sundari, who was a women trapeze artist and animal tamer in the Great Bengal Circus. The plot gives fictional currency to how women performers and their roles in circus were understood and the patriarchal frames through which her circus labour was constructed and represented.

Works Cited

Anderson, Benedict. *Imagined Communities: Reflections on the Origin and Spread of Nationalism*. Verso, 1991.

Artaud, Antonin. *The Theatre and its Double*. Translated by Mary Caroline Richards, Grove, 1994.

Bandopadhyay, Shirsho. *Shardulsundari.* Kolkata, Ananda, 2016.

Basu, Debshis, editor. *Professor Bose er apurbo bhromon brittanto (The wondrous travels of Professor Bose).* Kolkata, Karigar, 2013.

Basu, Sri Abanindra Krishna. *Bangalir sarkas* (*Bengali circus*). Kolkata, Gangchil, 2013.

Bhabha, Homi K. "Of Mimicry and Man: The Ambivalence of Colonial Discourse". *Discipleship: A Special Issue on Psychoanalysis*, vol. 28, 1984, pp. 125–33.

Bhattacharya Nikhilesh and Deeptanil Ray, editors. *Astracharcha: Collected Works of Pulin Behari Das (1877–1949).* Jadavpur UP, 2015.

Bouissac, Paul. *Circus as Multimodal Discourse: Performance, Meaning and Ritual.* Bloomsbury, 2014.

———. *Semiotics at the Circus: Semiotics, Communication and Cognition.* De Gruyter, 2010.

Chattopadhyay, Bankim Chandra. *Anandamath (The abbey of bliss).* Ramanujan UP, 1882.

Damasio, Antonio. *The Feeling of What Happens: Body and Emotion in the Making of Consciousness*. Vintage, 2000.

De, Sarmishtha. *Marginal Europeans in Colonial India: 1860-1920.* Kolkata, Thema, 2008.

Dutt, H. *Lieut. Suresh Biswas: His Life and Adventures*. Kolkata, P.C. Dass, 1899.

Figlerowicz, Marta. "Affect Theory Dossier: An Introduction". *Qui Parle*, vol. 20, no. 2, 2012, pp. 3–18.

Foucault, Michel. *Discipline and Punish: The Birth of the Prison*. Translated by Alan Sheridan, 3rd ed., Vintage, 1995.

Ghosh, Anirban. *The Tropic Trapeze: Circus in Colonial India*. 2014. Ludwig-Maximilians-Universität München. PhD dissertation.

Gregg Melissa, and Gregory J. Seigworth. *The Affect Theory Reader*. Duke UP, 2010.

Macaulay, Thomas Babington. *Macaulay's Essay on Warren Hastings*, edited by Margaret J. Frick, Macmillan, 1900.

Massumi, Brian. *The Politics of Affect*. Polity, 2015.

Merleau-Ponty, Maurice. *The Visible and the Invisible*. Northwestern UP, 1968.

Mukhopadhyay, Bhudeb. *Swapnalabdha Bharater itihas* (*The history of India as received in a dream*) Kolkata, 1895.

Orozco Lourdes, and Jennifer Parker-Starbuck, editors. *Performing Animality: Animals in Performance Practices*. Palgrave Macmillan, 2015.

Rosselli, John. "The Self-image of Effeteness: Physical Education and Nationalism in Nineteenth-century Bengal". *Past and Present*, vol. 86, no. 1, 1980, pp. 121–48.

Shapiro, Lawrence, editor. *The Routledge Handbook of Embodied Cognition*. Routledge, 2017.

Tait, Peta. *Wild and Dangerous Performances: Animals, Emotions, Circus*. Palgrave Macmillan, 2011.

THIRTEEN

IMPERIAL MALADY

EMPIRE AND AFFECT IN COLONIAL NARRATIVES

Anuparna Mukherjee

> Now I'm sick to go 'Ome—go 'Ome—go 'Ome!. . . .I'm sick for London again; sick for the sounds of 'er, an' the sights of 'er, and the stinks of 'er; orange-peel and hasphalte an' gas comin' in over Vaux'all Bridge.
>
> 'The Madness of Private Ortheris' (217)

The everyday ordeals of suffering inclement weather, heat, dust and tropical rain in the deltaic lowlands breeding manifold diseases, engendered a vast corpus of literary outputs consisting of medical literature, memoirs, diaries, fiction, journal entries, travelogues, manuals and *vede mecums* (handbooks or guides) in British India. They envisaged the empire as a sick body infested with deadly and 'putrefying ailments'. The visceral experience of wilting, weathering and shrivelling—corresponding to high rates of mortality and disease—brought forth spells of desperate to morbid nostalgia for 'home' among the colonial settlers as an 'affective aftermath' of their daily condition. There are references aplenty to wizened, consumptive and diminished bodies with images of corporeal decay in these imperial narratives. This chapter locates nostalgia at the intersection of ecological and medical discourses to explicate the affective contours of colonial modernity variously associated with notions of palliative care and remedy.

Pratik Chakrabarti, in *Medicine and Empire*, averred that mortality was exceptionally high among the soldiers who contributed invaluably towards the exponential growth of the British empire around the world. More people in the army died from sickness, malnutrition, bodily abuse, and exhaustion in inhospitable climes than those from actual injuries suffered during battle. Consequently, homesickness was of critical concern in the military and naval outposts of imperial regiments as one would see in Rudyard Kipling's 'The Madness of Private Ortheris' (published first in the *Civil and Military Gazette* and later anthologised in *The Plain Tales from the Hills*). Through a discussion of these texts, I will address nostalgia in its affective complexities that impacted the ecology of the empire through colonial narratives branching into different disciplines. The textual constructions of nostalgia and the attendant affects upon homesick Europeans will open up the ways in which it facilitated the growth of 'Little Londons' or customised 'Pocket-sized Europes' in locales outside the British heartlands.

Here, I primarily engage with fictional literature, such as Kipling's *The City of Dreadful Nights* and *Plain Tales from the Hills*, which record recurrent cultural/biological anxieties while recounting imperial experiences in India. From a literary vantage point, this chapter reads affect in relation to the transition of pre-modern urban ecology into colonial modernity, through the transformation of embodied spaces and the concomitant attempts to modify 'local' environs in the project of imperialism. As Jane M. Jacobs succinctly posits in the *Edge of Empire,* the globality of imperialism equally rests on the 'micro-management' and ordering of the local: 'The embeddedness of imperialist ideologies and practices is not simply an issue of society or culture but also, fundamentally, of place. This can be seen in the way in which racialised constructions are, often quite literally, made through place' (40). In the colonial narrative, the local becomes an embattled site of control through the inscription of identities, affects and affiliations in various contested formations and tangled social, ideological and cultural constructions of 'place'.

NOSTALGIA AND ITS AFFECTIVE HISTORIES

Nostalgia, as 'homesickness', was earlier touted as an illness affecting the body and mind. The inclusion of nostalgia in the medical vocabulary of the seventeenth century, in light of the emerging discourse on anti-nativity, deserves scrutiny to comprehend its moorings in colonial literature. There are tactile, visual, gustatory or other sensory and corporeal experiences that evoke nostalgia which bears a strong embodied dimension. Etymologically, in the *algos* of nostalgia, there is a clear allusion to an ailment of the body and the mind. When '-algia' is suffixed to a word it denotes a state of pain, such as neuralgia (nerve pain), myalgia (muscle pain) and so on. Indeed, when Johannes Hofer forged the modern coinage by cobbling together *algos* with the Greek *nostos* (tentatively 'homecoming') for his *Basel Dissertation* (1688), he ostensibly marked it as a 'cerebral disease' among Swiss students, professionals and mercenaries in the unfamiliar terrains of France, Germany and Italy who were woefully languishing away in their desire to return to their faraway homes. The memories were affected to such a degree that they muddled temporal trajectories and complained of hearing hallucinatory voices. The lovelorn yearning of nostalgia in Homer was thus reformulated as a malady with visible symptoms that tormented those bodies that did not belong to the place.

The *Cyclopaedia of Practical Medicine Vol. II* (1833) affirmed that the affliction of nostalgia 'has its source in the very frame and constitution of human nature in every part of the world' (53), thereby expanding its contours to adumbrate other cultures and experiences of dislocation. The manifestations of nostalgia included a gamut of psychosomatic conditions such as a compulsive thinking about home that overwhelmed and obscured other concerns, strong bouts of anxiety, irregular heartbeat, gastronomical disorders, insomnia, fever and frenzied outbursts. An entry, 'Nostalgia: A Vanished Disease', from *The British Medical Journal* in April 1976 made a comprehensive catalogue of these symptoms which 'we would mundanely call an anxiety state or a reactive depression; but some of them—stupor, fever, night sweats, delirium, hallucinations—point with hindsight to other diagnoses. So too, does the criminal behaviour, including arson' (857). In more serious cases, nostalgia led to self-harm, laceration and even insanity. Between the seventeenth and the twentieth centuries, psychiatric hospitals listed nostalgic affect as among the causes of mental perturbation along with religious excitement, love and seduction or domestic turbulence.

There are copious examples in the literature of the eighteenth and nineteenth centuries of moping characters who waned and withered away in utter despondency by nostalgically pining for their homeland in vain. While mapping the early history of the affect, Lisa O' Sullivan notes how nostalgia was categorised as a 'disorder of desire' (192) like 'lovesickness' that indicated the 'pathological potential of excessive emotion' (192). It was thought that the 'impact of emotion on the body took place through the mediation of imagination, whose role in insanity and disease had been well established in European medical thought by the 15th or 16th century' (192) through the clinical theory of 'humours' which regulated the activities of the body and the mind.

The British literature of the Raj is swamped with examples of corporeal suffering and the associative fear of dropping dead in a distant land that 'glisten[s] with eternal heat' (24).[1] Such painful thoughts often aggravated the psychic turmoil, creating a suitable environment for a burst of nostalgia. In his long poem 'The City of Palaces', James Atkinson writes about how European colonists constantly looked homeward for 'milder suns on happier seasons':

> Away, vain thoughts! The blazing sky proclaims
> Another clime. Here we must groan and sigh
> Under oppressive heat . . . (Stanza X, 10)

These contagious spells of nostalgia proved particularly troubling during the military expeditions around the world. From the American Civil War to the French Conquest of Algiers soldiers succumbed to virulent attacks of homesickness, the morale of the army dwindled and the risk of desertion increased manifold. The symptoms of nostalgia in a soldier were dealt with chastisement and threats of punishment, as it was seen as a moral flaw. Such signs of frailty created further embarrassments in colonial regiments, since the 'empire was above all a massive assertion of masculine energies' (Tosh) and fortitude. Yet in the military barracks, the soldiers who fought in difficult and remote places to secure Britain's imperial ambitions often yielded to homesickness due to their torturous lives, isolation and distance from home. Rudyard Kipling in 'The Madness of Private Ortheris' narrates how the young eponymous character breaks down under the afflictions of nostalgia, especially when in a state of drunken torpor. He descends into 'fits' of madness in his desperation and raves about going to Bombay to board a return ship.[2] Invoking his 'ome-sickness' (216), he deliriously pleads: 'I'm sick o' this dorg's life . . . Let' me go!' (218).

The misery caused by spatial dislocation was often intensified by routine rounds of 'seasoning sicknesses' and other environmentally induced ailments that led grieving soldiers like Stanley Ortheris to gripe and moan for their 'unreachable' home in England, which looked ever more beautiful in nostalgia's ineffable afterglow: 'If I 'ad a stayed of 'Ome, I might a married that gal and a kep' a little shorp in the 'Ammersmith 'Igh.—S. Orth'ris, Prac-ti-cal Taxidermist' (215). Here, determined to act upon his desires, Ortheris even risks the shame of desertion: 'I don't care. It's all one to me. *'Ow d'you know I ain't 'fraid o' dyin' 'fore I gets my discharge paipers?'* (216, italics mine).

In *Nostalgia in Transition, 1780-1917*, Linda Austin narrates a similar situation for charting the radical realignment of nostalgia from illness to an 'aesthetics of simulation' (4) in order to manage the disaffected soldiers reeling from 'homesickness' in military and naval bases. Often, these cadets were not transported back to their homes. Instead, attempts were made to rejuvenate them by recreating, in their workplaces, some of the things that the afflicted patients enjoyed in their native dwellings. Austen invokes Friedrich Schiller's treatment of a nostalgic patient, Grammont, in his early career as a medical practitioner. He improvised recreational exercises in the countryside, arranging visits to the spa or brief escape into the pastoral, to replicate some of the activities Grammont could have undertaken back home as a substitute for his actual return. It is largely this logic of similitude and simulation as a way of coping with the chaos of colonial life that drove the 'homesick' settlers to construct their 'Pocket-sized' Europes or portable Londons in the peripheries of the empire. However, here, nostalgia's aesthetic and palliative roles also intersect with the political and the social within the larger framework of imperialism through the creation of new affective ecologies in the colony.

The Home and the Empire

After the Industrial Revolution, home, and the familial values it stood for, simultaneously constituted the core of the bourgeois moral world and a refuge for the working class from the coercive labour in the marketplace. John Tosh, in *Manliness and Masculinities*, alludes to the maxim, 'an Englishman's home is his castle' to explain how it conveys a 'double meaning of possession against all comers, and of refuge or retreat from the world beyond' (Chapter 2, Section III).

Tosh opines that 'this second meaning spoke with special force to those middle-class men who experienced the world of work as alienating or morally undermining' (Chapter 2, Section III). The fact that the home was cardinal to the political and cultural praxis of imperialism can be partly understood from the command it had over its occupants since their very childhood. Britain's overseas subjects were anxious to create a home, however provisional, with ideas imported from their native country to their adopted habitats. However, one can hardly ignore the skewed contradictions in the spatial construction which homesickness engenders in colonial narratives. On the one hand, by expelling the dark and dangerous from domestic settings to far-off colonies through the recurrent evocation of a diseased landscape—ranging from mortiferous, frowzy wilderness infested with deadly pathogens to uncongenial climate and vegetation—nostalgia constructs a sanitised version of home in the collective imaginary. On the other hand, nostalgic affect commands the imposition of a model of familiarity to bring into practice that much-romanticised English domesticity in radically alien spaces. Since nostalgia in its prosthetic role thrives on replication, I shall explore how the British community in India attempted to recast their 'home' and coetaneous memories by reconfiguring the embodied experience of exile through the incorporation of forms and the conventions of sociality, grafted from their native land. Thus, in the combination of factors that led to the growth and transformation of lived spaces in British India, nostalgia, though often underplayed, had a role in discerning affective practices, attitudes and values.

Beyond its specific clinical manifestations, nostalgia clung to minds as an affliction, generating a complex of affects which Heather Houser describes as a set of 'body-based feelings that arise in response to elicitors as varied as interpersonal and institutional relations, aesthetic experience, ideas, sensations and material conditions in one's environment ... The feeling grounds one in the present, but it is also coded by past experience and impinge on the future' (3). The affective history of the empire, then, invokes the embodied responses and 'representation that both takes affect as its object and attempts to elicit affect' (Agnew 301). It explores the relationality between everyday 'experiences' feelings and bodily responses to the environment (some of which were held in common) in interpreting important administrative decisions in the colonies, rather than the empirical structure of the unfolding events.

THE HILLS AND THE PLAINS

Let the *City Charnock* pitched on—evil day!
Go Her way.
Though the argosies of Asia at Her doors
Heap their stores,
Though Her enterprise and energy secure
Income sure,
Though "out-station orders punctually obeyed"
Swell Her trade—
Still, for rule, administration, and the rest,
Simla's best.

'A Tale of Two Cities' (Kipling 82)

Kipling, like many of his compatriots, shared a visible dislike for the soggy weather in the plains and was particularly hostile to the polluted city of Calcutta, the administrative and commercial capital of British India before 1911. He reserved the most underwhelming sobriquets for the colonial headquarter calling it the festering city of 'cholera', 'cyclones' and 'crows' that feed on garbage and scrapheaps. In *The City of Dreadful Nights* he complains of a vile, gut-wrenching smell enveloping the urban landscape—the 'Big Calcutta Stink'—'that blackens the face of any Englishman who sniffs it' (18). Kipling identifies it as the 'reek of Calcutta', notorious for its 'diffused, soul-sickening expansiveness' (8), issuing from the untreated refuse that was responsible for multiple illnesses and epidemic outbreaks under colonial administration.

Cholera was one in an inventory of fatal diseases that the Europeans feared in the tropics. Its germs allegedly drifted 'through the muggy atmosphere, and stuck in the branches of trees like wool-flakes' (Kipling, 'A Germ Destroyer' 101). Believed to be exacerbated by the heat and the humidity of the riverine plains, the cholera epidemic—which posed much anxiety for doctors and public health practitioners in the nineteenth century—took hundreds of lives every year especially in the monsoon season. The medical sciences produced various postulations and 'germ theories' on bacteria and microorganisms. In fact, Kipling's short story, 'A Germ Destroyer' narrates a debacle caused by the experiment of an eccentric scientist who, after extensive research in lower Bengal, (101) invented an 'Invisible Fumigatory' (101) for eradicating cholera.

Then, of course, there were other febrile ailments, such as malaria, caused by the bites of vicious mosquitos from the swamps. In *Hartly House, Calcutta,* the female protagonist recalls with great consternation: '. . . muskettos; I must tell you, though I shudder at the bare recollection of so vulgar a nuisance, that, in like manner with the bugs in London, they mercilessly annoy all new-comers, blistering them, and teasing, if not torturing them continually; and in a great measure spare those who are seasoned to the climate' (Gibbs 10).

In the poem 'The Mosquito's Song', a Company poet John Young (1797–1846) iterates similar horror of living in the plains. But instead of putting forth the human point of view, he uses the mosquito to narrate its story:

> Oh the pleasures of the plains
> In Bengal, and in the Rains,
> When the climate, damp and warm,
> Makes our tiny tribes to swarm

From each puddle, from each tank,
Fringed with vegetation rank; (56)

Then it goes on to declare its delight in relishing the 'unsuspecting' Griffin's blood, when on a lazy summer day, 'flushed with heat' (57), the Company servant dozes off in a noontime siesta, dreaming of his home and the 'youthful village love' (57) he had left behind to 'rove' in a foreign land. Like other poems marking the pitiful life of colonials in the backwaters of the empire (and the delight of the mosquitos, glowing in the 'pleasures of the plains'!), this too juxtaposes the afflictions of the present with the joys of the past. It only heightens the discontent using contrastive registers of bodily gratification and pain, such as 'sucking', 'swilling' or filling the 'reddening bodies' (of the mosquitoes) with blood, and so on. Likewise, the affective expressions of misery are sharply physical, ranging from mild discomfiture to a more chronic and visceral pain that speaks of 'festering sore', to the sheer inability to sleep or rest: 'Sleep no more!/Gnats have murdered sleep.'[3] Here, the incremental irritation caused by the mosquito bite builds up the tension for the accursed Griffin in the poem:

He sits up
Startled, —scarce awake, —head bursting, —
—*Itching*, —*scratching*, —*smarting*, —*thirsting*; —
Curses deep, and loud, and long,
Yet unsated, chaunt their song.
Oh the pleasures of the plains
In Bengal, and in the Rains! (57–58)

Kipling becomes one in this extensive line of European writers in British India, who cites Calcutta's uninhabitable condition to advocate shifting the capital to the unspoilt hills of Shimla in North India. Kipling specifically uses 'filth' to make a series of moral judgements on Indians coterminous with their lack of civic sense. This coheres with the spatial discourse of colonialism that brings together morality, muck and disease to prop up the narrative of a civilising mission by conflating a sick landscape with weakness of the body and character. In his harangue against the city, Kipling affirms that 'The Big Calcutta Stink' 'resembles the essence of corruption' (9) and blames it on the conspicuous presence and interference of the native babus in the Calcutta Municipal Board: 'The damp, drainage-soaked soil is sick with the teeming life of a hundred years, and the Municipal Board list is choked with the names of the natives—*men of the breed born in and raised of this surfeited muck-heap*!' (*The City of Dreadful Nights* 10; italics mine).

However, Calcutta's urban ecology was, in a sense, no different from the big industrial cities in Europe. In London 1858, the phenomena that came to be known as the 'Great Stink' resulted from noxious vapours rising from the Thames. A faulty sewerage system that led to the accumulation of human excreta and industrial wastes along the river lines caused the spread of fatal or highly contagious diseases in the heart of the metropole. Similarly, David Barnes, in *The Great Stink of Paris and the Nineteenth-Century Struggle against Filth and Germs,* extensively writes about the 'foul odors of Paris' (1) that covered the city and the medical hazards it entailed:

> In the late summer of 1880 in Paris, death was in the air, and it smelled like excrement. That, at least, was the prevailing opinion at the time, shared and vociferously proclaimed by scientists, medical doctors, elected representatives, and ordinary Parisians. For more than two months, oppressive and insufferable odors pervaded the air of the capital, occasionally disappearing and then reappearing with even greater intensity according to the indecipherable rhythm of mysterious forces (1).

Nevertheless, the deleterious effects of the tropical climes upon the European body, couched in incessant complaints about their lives in the lower Gangetic plains, grounded the desire of finding an alternative sanctuary in higher altitudes closer to the environment of the home. Both pathological and lyrical outpourings of nostalgia for the rolling hills 'with their mild climate', which backed this desire of recreating 'an England in the Tropics' (Pradhan 34), underpin the colonial interventions in transforming the ecology of the Indian Empire. The affective reactions to the multifarious physical distresses and vexations that the perspiring, parched body suffered in the heat and dust of the lowlands, coupled with the disturbing mortality rate among Europeans, expedited the birth of about eighty hill stations modelled on English towns between 1820 to 1880. Here, the wholesome, balmy mountain air and unsullied surroundings once again brightened up the possibility of reviving memories of home. The southern hills of the Nilgiris were deemed particularly suitable for this project due to their temperate climate and moderate rainfall in the valleys: 'Here one could find landscapes to feed the most voracious nostalgia. Ootacamund, known as Ooty by the "abbreviating Saxon", was most commonly likened to the South Downs of Sussex. Many were the other candidates, however:

'"Hertfordshire lanes, Devonshire Downs, Westmoreland lakes, Scotch trout streams and Lusitanian views" (Lord Lytton) . . .' (Herbert 121).

Importantly, the birth of these hill stations received impetus from the spurt of Western science's new-fangled interest 'in epidemiology and the health-restoring benefits of higher altitudes' (Herbert 99). In midst of the mountains, as Eugenia Herbert notes in *Flora's Empire: British Gardens in India,* 'a very different India beckoned, an India where familiar flowers bloomed, where nights were cool and sallow complexions regained a rosy glow. One could almost imagine one was at home' (99). Consequently, extensive plans were rolled out to set up sanatoriums in places like Darjeeling. In the 1830s, Governor General Lord William Bentinck deputed General Lloyd to negotiate with the Rajah of Sikkim for the possession of Darjeeling, in exchange for an equivalent sum of money or land. A deed dated 1 February 1835, finalising the transfer, stated, 'The Governor General, having expressed his desire for possession of the Hill of Darjeeling on account of its *cool climate, for the purpose of enabling the servants of his Government, suffering from sickness,* to avail themselves of its advantages, the Sikkimputtee Raja, out of friendship for the said Governor General, hereby present Darjeeling to the East India Co. . . .' (Aitchison 325, italics mine)

These mountainous retreats that had spatial and climatic similarities with the English countryside became vital in the new discourse of tropical medicine, which explored various palliative 'experimental methods of cure and prevention' (Ghosh 43) of diseases in the colonies. Several restorative institutions, ranging from sanatoriums and 'convalescence depots' to infirmaries and special hospitals, were established in these places to nurse the sick administrative servants of the Raj amidst the healing landscape. Such ameliorative medical reforms were founded on the same logic of 'semblance' and 'simulation' as invoked in Schiller's treatment of his nostalgic patient, Grammont. We find corroboration of the sentiment behind such measures in Waltraud Ernst's seminal work on colonial psychiatry, *Mad Tales from the Raj,* where he purports that 'there was no better cure for an ailing and alien Company servant than "home". This view, cherished by most Europeans in British India, was shared by the majority of the medical profession' (123). And indeed, away from the malarial plains of the 'down country' a retreat in the mountains covered with snow and 'familiar plants and trees' immediately 'evoked wave after wave of delicious nostalgia' (Herbert 107). However, nostalgia's affective labour of resurrecting the environment of the home worked towards expanding the colonial geography by homogenising the landscape through repetition, to create narratives of continuity, organisation and unity that would envisage the empire as 'a part of an imperial whole' (Auerbach 47).

The spatial configuration invoked by Kipling in the title of his anthology of short stories— *Plain Tales from the Hills*—is based on the construction of the hills as belonging to the colonisers from where they would rule over the vast subcontinent, observing,[4] 'surveying, archiving and inventing narratives' (Nagai xxv). Hills were touted as 'empty landscapes waiting to be appropriated by the "settlers", unimpeded and uncontested' (Pradhan 34) by removing the memories of forced acquisition or deeply fraught negotiations with the restive native powers.

Conclusion

The curative discourse attending to nostalgia in the nineteenth century conflated the art of simulation with pharmaceutical intervention, which was entangled in larger environmental debates about the new spatial production of these colonial and commercial enclaves in the hills. The implications extended to encompass various colonial endeavours from the initiation of the plantation economy, medical tourism, to the modification of local vegetation to replicate those in the English villages. The ecological and cultural transformation of these spaces then clearly facilitated the imperial propaganda of embedding its political footprints deeper into 'foreign' soil, through the dissemination of its knowledge, conventions and practices. The European landscaping of the hill stations or what Herbert calls, Britain's 'garden imperialism'—that is, the eagerness to recreate European plants in colonial gardens—anchored nostalgia in a chain of insidious desires which surreptitiously aspired to control the 'untamed' wilderness of India with a stamp of its own order and civilisation. In his introduction to *Warm Climates and Western Medicine*, David Arnold asserts that the framing of 'tropicality' to understand the afflictions of Europeans in lands 'geographically distant and culturally remote' (6) was itself an imperial manoeuvre. It fostered a perception of nature that straddled the extremities between fecund and beautiful to dark and deceitful, traits that were very often attributed to the natives of these lands.

Thus, the elaborate exercise of identifying, classifying, demarcating and disciplining and colonial landscape into 'readable spaces' (De Certeau 36) was an essential extension of the 'civilising mission' that created a subterfuge for the legitimisation of colonial exploitation and conquest.

These moves were largely perpetuated by the visual representation of the empire by the commissioned artists of the Raj through numerous sketches, paintings and aquatints—especially those belonging to the 'picturesque' tradition—popularised by William Gilpin in 1782. The fabled landscape of the Orient was defamiliarised as strange, sublime and dangerous, and yet with deeper affective engagement, one could discover an organic connection obscured by time.

In his *Journal,* the early European artist, James Baillie Fraser draws explicitly on the visual connection between home and the empire. This romanticised nostalgia for the 'lost' other was predicated on a strong urge towards reclamation and consumption.

> Asia was almost lost in our imagination: a native of any part of the British Isles might here have believed himself wandering among the lovely and romantic scenes of his own country. The delight of such association of feeling can only be understood by those, who have lingered out a long term of expatriation, and who anxiously desire the moment of reunion with their native land. (107–108)

Yet, these hills were never entirely 'English' even when harnessed and Anglicised. And the fantasised space of 'whiteness' was pervaded by the natives. This disjuncture 'between the mother culture and its bastard' (Bhabha 153)—that was at the crux of the nostalgia experienced by the expatriates relentlessly hankering for both the English climate and heritage—also fostered sites of ambivalence integral to any instance of colonial mimicry. It is this 'slippage'—dissimilarity or inadequacy—which played up as a potential tool for subverting neat spatial demarcations by both disturbing and corrupting the 'visibility of the colonial presence' (Bhabha 154) in the empire.

To sum up, nostalgia as a symptom of colonial modernity was a shaping force in interpreting the ecology of the colonies in a specific way in order to anchor anomalous bodies in spaces that were both perilous and alien. However, the politic/poetics of nostalgia imbricated in multiple discourses of the time contributed to the imperialist propaganda as we see in the ecological *vis-á-vis* the medical mapping of affect through the literary framework, evoked in this chapter. The deployment of nostalgia for the 'past environment' that served as a template for envisaging local ecologies of newly discovered places through comparison and contrast was formative in eking out the settlers' identity in the slippery terrain of the empires. However, this environmental imagination, that simultaneously created the 'exotic paradise' and then dismantled it with a model of familiarity on occasions when it was deemed necessary, was an exploitative instrument in the colonial machinery. Such ecological nostalgia in colonial narratives often came with the desire to modify, if not erase it altogether, other memories and lifeworlds which nonetheless survived as resilient traces from 'different' times.

Notes

1. 'Calcutta: A Poem', (1811) anthologised in *Poets of the John Company*, edited by Theodore Douglas Dunn.
2. Present day, Mumbai.
3. Here's an obvious echo of Macbeth's insomnia after his assassination of Duncan: Macbeth shall 'Sleep no more!/Macbeth does murder sleep'. These lines immediately evoke a visceral image of blood and pain.
4. The high altitude of the hills provided a good strategic location for surveillance and control over the surrounding territories, including the neighbouring kingdoms.

Works Cited

Agnew, Vanessa. "History's Affective Turn: Historical Reenactment and its Work in the Present". *Rethinking History*, vol. 11, no. 3, 2007, pp. 99–312.

Aitchison, C.U. "Deed Executed by the Raja of Sikkim Ceding Darjeeling to the English, dated 1st February 1835". *Collection of Treaties, Engagements and Sanads: Relating to India and Neighbouring Countries,* vol. 2, 4th ed., Superintendent Government Printing, 1909, pp. 325.

Anonymous. "Calcutta: A Poem". *Poets of John Company*. Selected and arranged by Theodore Douglas Dunn, Calcutta, Thacker, Spink and Co., 1921, pp. 21–24.

Anker, Peder. *Imperial Ecology: Environmental Order in the British Empire, 1895–1945*. Harvard UP, 2001.

Arnold, David. *Warm Climates and Western Medicine: The Emergence of Tropical Medicine, 1500–1900*. Rodopi, 1996.

Atkinson, James. *The City of Palaces: A Fragment and Other Poems.* Government Gazette P, 1824.

Auerbach, Jeffrey. "The Picturesque and the Homogenisation of Empire". *The British Art Journal*, vol. 1, 2004, pp. 47–54.

Austin, Linda. M. *Nostalgia in Transition, 1780–1907.* U of Virginia P, 2007.

Barnes, David S. *The Great Stink of Paris and the Nineteenth-Century Struggle Against Filth and Germs.* John Hopkins UP, 2006.

Bhaba, Homi. "Signs Taken for Wonders: Questions of Ambivalence and Authority under a Tree Outside Delhi, May 1817". *Critical Inquiry*, vol. 12, no. 1, 1985, pp. 144–65.

Chakrabarti, Pratik. *Medicine and Empire.* Palgrave, 2014.

De Certeau, Michel. *The Practice of Everyday Life*, vol. 1. U of California P, 1984.

"Editorial: Nostalgia: A Vanished Disease". *British Medical Journal*, vol. 1, no. 6014, 1976, pp. 857–58.

Ernst, Waltraud. *Mad Tales from the Raj: Colonial Psychiatry in South Asia, 1800–58.* Anthem, 2010.

Fraser, James Baillie. *Journal of a Tour through Part of the Snowy Range of the Himala Mountains, and to the Sources of the Rivers Jumna and Ganges.* Rodwell and Martin, 1820.

Gibbs, Phebe. *Hartly House, Calcutta.* Edited with an introduction by Michael J. Franklin, Oxford UP, 2007.

Hancock, Thomas. "Nostalgia". *Cyclopaedia of Practical Medicine: Comprising Treatises on the Nature and Treatment of Diseases, Materia Medica and Therapeutics, Medical Jurisprudence, etc., etc., edited* by John Forbes, John Conolly and Alexander Tweedie, vol. 2, 1833, p. 53.

Herbert, Eugenia W. *Flora's Empire: British Gardens in India.* U of Pennsylvania P, 2011.

Hofer, Johannes. "Medical Dissertation on Nostalgia". 1688. Translated by C.K. Anspach, *Bulletin of the History of Medicine*, vol. 2, 1934, pp. 376–91.

Houser, Heather. *Ecosickness in Contemporary U.S. Fiction: Environment and Affect.* Columbia UP, 2014.

Jacobs, Jane M. *Edge of Empire: Postcolonialism and the City.* Routledge, 1996.

Kipling, Rudyard. "A Tale of Two Cities". *The Collected Poems of Rudyard Kipling.* Introduction and notes by R.T. Jones, Wordsworth, 2001, pp. 80–82.

———. *The City of Dreadful Nights and other Sketches.* Alex Grosset & Co., 1899.

———. "The Germ Destroyer". *Plain Tales from the Hills.* Penguin, 2011, pp. 100–04.

———. "The Madness of Private Ortheris". *Plain Tales from the Hills.* Penguin, 2011, pp. 214–21.

Nagai, Kaori. Introduction. *Plain Tales from the Hills.* Penguin, 2011, pp. xiv–xxxvii.

O'Sullivan, Lisa. "Lost Imagination: French Nostalgia and the Turn to Memory". *Memory Studies*, vol. 3, no. 3, 2010, pp. 192–95.

Pradhan, Queeny. "Empire in the Hills: The Making of Hill Stations in Colonial India". *Studies in History*, vol. 23, no. 1, 2007, pp. 33–91.

Tosh John. *Manliness and Masculinities in Nineteenth-Century Britain.* Routledge, 2016.

Young, John. "The Mosquito's Song". *Poets of John Company*. Selected and arranged by Theodore Douglas Dunn, Calcutta, Thacker, Spink and Co., 1921, pp. 56–58.

FOURTEEN

BETWEEN SPACE AND THE BODY

THE AFFECTIVE ECONOMY OF CASTE

Antaripa Bharali

Hannah Arendt, in *The Human Condition* (1988), states that 'no activity can become excellent if the world does not provide a proper space for its exercise' (49). Defending the common world that we share as finite individuals, but which continues to live on even after us, Arendt stresses our human capacities to make the world more permanent than us through the activities that we undertake in it. However, this capacity to make, to begin, to create is possible only when 'each human activity has its proper location in the world' (Arendt 73)—that is, provided with its natural space to actualise itself. Following this forceful claim, this chapter attempts to ask what happens when this space is denied to certain individuals. What role does space play in prohibiting us from imagining a common world for all if it makes available freedom for some and unfreedom for others? Reflecting on such questions at a case-specific level, this chapter attempts to think through the role of space and its enabling/disenabling capacity in a caste society. I interrogate the determinate ways in which space (within the context of caste) regulates the untouchable body. I argue that space provides an enabling condition for a particular affective experience of the untouchable body, which in its most direct manifestation, is sensory. However, the body too—being a site of concentrated valuation—produces and sustains relations of proximity and/or distance that generates social boundaries in a given space.

This chapter endeavours to explicate how space both reproduces relations of caste and makes explicit the implicit socioreligious sanctions embedded in a caste society. The first two sections do so by looking at how space is organised through caste in favour of some and to the disadvantage of others, and how spatial metaphors have a role in disciplining a caste-inscribed body either through direct violence or by regulating our sensory experience and their perception in the given ecology. Building on the latter point, the last section adds a phenomenological dimension to the reading of space and highlights the linkages between the concepts of space, caste, and body.

SPACE AS AN ENABLER: CASTE, EXPERIENCE AND CONCEPT

Gopal Guru's essay 'Experience, Space, and Justice' in *The Cracked Mirror* (2012) offers a discussion of the enabling role that space plays in making explicit the implicit tendencies of caste and its operations in our society. Guru develops at various places a very insidious but often ignored relationship between experience, space and concept. However, before that relationship is explicated it is important for our purposes to lay clear, albeit briefly, the theoretical underpinnings of the idea of space. What does it do? How is it structured? How does it function? What effects can it produce? Foucault and Lefebvre are key thinkers who have addressed these questions, especially those surrounding the production and organisation of space, which can provide a preliminary understanding of how space is theorised in popular discourse.

For Foucault, the purpose of any form of spatialisation is to partition space or the object in a way that it can be analysed more easily or more directly, and so that it is exposed to cognitive processes in a more detailed manner. This produces scientific knowledge, the consequences of which are manifestly 'disciplinary' (Foucault 52). Lefebvre, similar to Foucault, looks at the way power and ideology makes use of space and extends the argument to perceptively ask, 'what is an ideology without a space to which it refers, a space which it describes, whose vocabulary and links it makes use of and whose code it embodies' (Lefebvre 44). As it becomes theoretically observable through these formulations, space is not a natural category but is always culturally and politically constructed and appropriated with far-reaching consequences for regimes of power and discipline.

All kinds of social praxis or relations depend on space as an enabling condition where space is structured, or the architecture organised in ways that facilitate specific modes of relationships and experiences in society. However, just the condition of 'being in a space' does not allow one to conceptually understand one's place in those specific relations since this kind of conceptualisation depends on how one experiences that particular space. And it is this connection between space and experience, which Gopal Guru pays special attention to in his essay, where experience introduces a certain dynamism into a hitherto static space, or in other words, brings it to life. In another sense, space or spatial organisations in turn hold the capacity to objectify a person based on how they experience a particular space. This happens as space—and for that matter even time—though abstract and continuous categories, are always identified and localised when experienced by an individual as a specific place or moment. This way of experiencing space results in a kind of objectification where now the individual is an object-in-space, inaugurating processes of value attribution where the objects are either clean or dirty, pure or impure, servile or dominant, and so on. Experience, which is 'subjectively realized but objectively produced through the logic of space' (Guru, "Experience, Space and Justice" 72), finds its theoretical representation through concepts that are mediated by this very experiential space. Gopal Guru uses Ambedkar and Gandhi as examples and shows how their respective experiences with space shaped their vocabularies differently. He writes that as Gandhi travels throughout the country through favourable social spaces such as public *maidans*, he 'ceases to be a bania (trading caste) or a Gujarati and becomes a Mahatma' and is able to take voyages and 'occupy the central spaces that he uses so tactfully for the political mobilization of the Indian masses' (103) against their colonial rulers through concepts like self-rule. Gandhi thus had a choice, Guru argues, to transgress spaces vertically. However, 'Ambedkar's social constituency only opened up horizontally, moving from one dalitvada to another' (104). Further instances include a judge talking to Ambedkar in a moving vehicle to avoid sharing the same space as him or Ambedkar's own teacher not being able to call him home, despite expressing such a wish, for fear of social repercussions. These experiences provided for Ambedkar the testimonial ground on which he chose, Guru argues, to prioritise the concept of self-respect over self-rule (unlike Gandhi) unveiling the relation that experience has to one's political vocabulary.

With regards to Ambedkar's imagination of the spaces around him, as a parenthesis here it is important to take note of his own vocabulary to mark out the different words that Ambedkar uses to describe the spaces within which he finds himself and his community. Words like 'ghetto' (Ambedkar, "Untouchables or The Children of India's Ghetto" 19) or 'dungeon' (Ambedkar, "Waiting for a Visa" 675) find multiple utterances within Ambedkar's thought, often while describing the sense of separation that untouchables in Hindu society are subjected to as a result of their physical segregation in space. These descriptions, which imply restriction in movement, raises the important question of movement and mobility of dalit bodies within space. Caste here functions in a way that makes space, both for touchables and untouchables, fixed and static (especially for the latter)—a move which is ensured through punitive measures that are taken when there is movement that would lead to violations of caste rules regarding touching.

What is the import of a space that disallows movement? How is space to be thought of or re-thought of in the absence of the ability to move within? Not only does space in such a context freeze itself, but it also produces significant effects such as feelings of isolation and loneliness that subjectively cripples an entire community. Caste, by freezing all points of contact or even its distant possibility by means of physical segregation, isolates and forces loneliness on a whole section of its members. Ambedkar discusses this precise effect of segregation as the 'problem of isolation' ("Untouchables or The Children of India's Ghetto" 112) where, in addition to describing the problem as an existential state that the community of untouchables are perpetually faced with, he exposes the ways in which the problem of untouchability finds no 'natural allies' (116) to ensure its annihilation because of this enforced isolation. This experience of isolation is brought into horrific clarity through one of Ambedkar's autobiographical episodes from 'Waiting for a Visa'. After returning from London and finding it impossible to find accommodation in Baroda as an untouchable, Ambedkar describes being forced to take refuge in a Parsi inn. Fearing being outed as an untouchable, he lives by impersonating a Parsi in a dark room without lights describing it as next to a hall 'filled with all sorts of rubbish' amidst the 'chirping and flying about of the bats' ("Waiting for a Visa" 675). Characterising himself as 'a single solitary individual', 'enveloped in [the] complete darkness' of loneliness and 'longing for the company of humans' (675), Ambedkar describes the time spent in that inn as being brutally alienating and exhibits an acute awareness of the misery and sadness that he is put though in the lonely 'dungeon' solely because of his caste identity. Commenting on this experience of one's shelter as a dungeon, Aishwary Kumar in *Radical Equality* (2015) raises the stakes of this narration arguing that through this comparison Ambedkar 'forces his readers to think the conditions of such life [sic]: a life constituted in its very alienation from other lives' (312)—that is, an untouchable life. This awareness of the problem of loneliness is made more visceral only as a result of isolation and separation in space. Space that is divided in ways that removes a section of people from contact, or more specifically from touching, serves as a perfect ground for 'visual proof of social hierarchy' (Duneier 16) where hierarchy and its immediate effects are produced and maintained not only through religious and social sanctioning but also through its experiential bringing to life.

Regarding the word 'ghetto', Ambedkar comments it is not simply a case of 'social separation, a mere stoppage of social intercourse for a temporary period. It (is) a case of territorial segregation and of a *cordon sanitaire* putting the impure people inside a barbed wire, into a sort of cage. Every Hindu village has a ghetto. The Hindus live in the village and the Untouchables in the ghetto' (Ambedkar, "The Untouchables" 266). It can be argued that the ghetto becomes a spatial necessity of any space organised according to caste relations. Social segregation and spatial exclusion are co-constitutive and limit every possibility of escaping such a structure without annihilating caste. In the title of Ambedkar's 'Waiting for a Visa', the word 'visa' can be taken to signify the desire to move away from caste, to escape its location and find oneself a space free from it. However, the difficulty of this possibility is immediately admitted with the preceding word 'waiting'. The term 'waiting' negates the easy possibility of this movement from one space to another and reminds one of the unending time of caste, its continued existence and its perpetual reality that one can only annihilate and not escape.

SPACE AND BODY: INTERRELATIONS

Space, through its enabling role, is able to intensify caste relations that are especially socially disadvantageous and morally crippling for one section as opposed to another. Owing to this, the preservation of one's space becomes intensely connected to preservation of one's caste. We may recall another example from Ambedkar's autobiography where he notes a case that appeared in Gandhi's journal 'Young India', where an untouchable schoolteacher fails to persuade a Hindu doctor to check on his wife who had just delivered. As a result of this, the wife and the child die for want of medical attention because the doctor does not want to cross over to the 'Harijan colony' (Ambedkar, "Waiting for a Visa" 687). Citing this, one can ask under what circumstances is such an exceptional case of cruelty made possible? Ambedkar observes here the fixity with which caste rules are followed, not only ritually but even spatially, such that the doctor in this case would rather be inhuman and flout his professional code of conduct than cross over and touch the untouchable or enter the 'Harijan colony' which itself has become untouchable now. This example demonstrates the peculiar case of how space itself comes to acquire the attributes of the body and becomes untouchable, as the doctor refuses not just to touch the dalit body but also the space of the dalit, that is, 'the dalitvada'.

This is made simultaneous with the body acquiring attributes of the space and being culturally marked, be it with a pot around one's neck (so that the spit doesn't fall on the ground and pollute it), or a broom around one's waist (so that one's footsteps are cleaned as one walks). In a manner similar to what was argued by the French theorist Georges Bataille, the body here becomes 'the ultimate measure of space and itself the space most fully occupied and most desirably penetrated' (qtd. in Cokal 76) and this 'writing on body' (Manjali 102) is what allows the brahminical system to turn dalit bodies into cultural sites ready for appropriation, and as a consequence, subject to violence. Recent scholarship on caste-inscribed bodies by authors like Aniket Jaaware, Prathama Banerjee and Y.T Vinayaraj have argued for a kind of 'nascent materialism' (Jaaware, "Eating, and Eating with the Dalit" 287) present within dalit thought that 'might prove to be more fruitful than a transplanted or translated leftist materialism' (288) in understanding its nature and politics. This materialism, unlike its counterpart, is centred around the category of the body wherein materiality is re-interpreted as 'a domain in which the human body becomes the locus of the operations of larger historical forces' and the body '[whether] starved, bonded, sick or violated becomes proof and product of material processes' (Banerjee 8).

In reducing one's fate to one's body where it becomes its most immediate reality, caste then mostly operates as practices through and on the body, be it the literal markings that are to be found on caste bodies (for instance, the sacred thread on the brahmin or the pot around the neck of the untouchable) or the refusal to touch the 'impure' body of another. The symbolism of the body is also evoked in the *Purusha sukta* which tells of the origin of the four varnas: where the brahmin emerges from the mouth of the Creator and the shudra emerges from the feet of the Creator, all parts imbued with extensive meaning providing a model for society to adopt. This model then allows, as Ambedkar observes, for 'fixing the functions of the four classes' and 'fixing the gradation of the four classes after a preconceived plan' ("Who were the Shudras?" 32) where, as a result of being placed at the level of the feet, the shudra is accordingly placed at the bottom of the caste hierarchy in society and 'is given the filthiest function, namely to serve as a menial' (33).

This close proximity between the physical body and society, that is, the social body at large, is not new. Mary Douglas, in her book *Natural Symbols* (1970), explores this link and argues that the physical and the social body continually exchange meanings, where all bodies through their 'appearance, movements and functions' (Ramanujam 43) are given cultural and socially determined meanings that saturate their physical presence. In a way, the body thus becomes a medium of expression of the rules, fears and anxieties of society. In the case of caste, such an imposition can be seen, for instance, in the concept of 'dirt' *vis-à-vis* the description of untouchable bodies. Here, dirt represents less a characteristic of an empirical reality and functions more as a concept that marks and demarcates boundaries between the pure/impure in the case of the physical body, and inside/outside in the case of the social body.

Douglas argues that instead of 'dirt' being a physical property of certain bodies, 'reflection on dirt involves reflection on the relation of order to disorder' (Ramanujam 44). So, anything that holds the capacity to threaten the inside and the pure is deemed 'dirt' as it represents disorder to the given order of caste. The anxieties of caste society to maintain its inside are transplanted on the body, where now the stability and the integrity of the social order of caste rests solely on maintaining the integrity and division between pure and impure bodies, both structures resting on the function of the body as a literal boundary. Ambedkar, in 'Untouchables or the Children of India's Ghetto' illustrates this point when he narrates an episode where an untouchable's body is used as a physical shield against an advancing procession to stop it from moving forward, as nobody dare cross or come close to it lest the members of the procession violate the known boundaries of their caste. One may observe here the curious case of the human body being used as a literal shield, as a marker of limit where now the prospect of transgression lies not simply outside of the body but is ontologically contained within the body and helps secure the boundaries that order caste society. It is important to remember here that 'limit', 'transgression', 'boundaries' are otherwise all qualities of space.

Such a link between the structures of the body and society are important insofar as they help us understand the relationships that ensue between the brahmin and the untouchable, where the former is now ridden with anxieties all the time to maintain one's purity in the face of impurities outside. These anxieties are best manifested by the various purificatory rituals that the brahmin undertakes every day to guard himself from the impurities of the outside world. Since (as mentioned) the notion of purity serves to structure the given order of caste while impurity tends to threaten that order, 'ritual purification (can be seen as) an act of restoring the given order' (Ramanujam 47). The body thus becomes an important reference point to study the workings of caste given the way that it is made to, not only imitate the social structure of caste society, but also to influence it.

Broadly arguing then, caste-based interactions with the dalit body occur along two primary axes. The first is that of untouchability and the other violence, that is, either the body is not touched or/and it is touched only violently. These two axes help us think of the body materially through touch which allows us to see the body in 'social terms and not merely as a unit of condensed matter' (Jaaware, *Practicing Caste* 23). Sundar Sarukkai, in his paper 'Phenomenology of Untouchability', distinguishes between contact and touch arguing that while the former is seen as a relational property between two objects coming into contact, the latter is seen as a 'quality that inheres in the object' (41). And it is because of this distinction, that 'untouch' becomes a property of the untouchable body—they remain untouchables whether or not they come into contact with another body.

However, such a theoretical manoeuvre of making untouchability the quality of a particular body is not properly understood without an assessment of the metaphysics of the body in general, and without its relation to the brahmin body in particular. Gopal Guru in response to Sarukkai's argument maintains that the Buddhist metaphysics of the body suggests that all bodies are 'a source of impurities' (Guru, "Archaeology of Untouchability" 50), be it in a ritualistic sense—insofar as it partakes in rituals of death, birth and others—and in the more material sense through its association with negative bodily properties like sweat, gas, and so on. However, this ontological equality between bodies is rejected in the case of caste where the impurities of some bodies are completely transferred onto others through a process of what Sarukkai calls supplementation and outsourcing. This hides the very fact that the property of untouch is actually not a property of the untouchable body but it is the brahmin body whose body is 'folded' (53) insofar as its touching abilities are muted—it is the brahmin body's touch which is disabled. Therefore, the brahmin body too assumes a matter of great importance when it comes to maintaining the caste order.

In fact, it is the brahmin body that is the most vulnerable when it comes to preserving untouchability. This vulnerability is manifested through the innumerable number of regulations that operate on their bodies, though mostly of a purificatory nature. The brahmin body therefore 'bears visual marks of its social position, just as the dalit body' (Jaaware, *Practicing Caste* 110). In *Practicing Caste*, Jaaware sees this vulnerability exhibited by the brahminical body to be a 'measure of the vulnerability also of the orthodox social order' of caste (110). The body again becomes a medium for the larger social order and is able to provide us with a more symptomatic reading of the society which we inhabit, where all the anxieties and the vulnerabilities of the community are expressed and marked onto the bodies of its members. Additionally, such processes of ritualisation and marking on the body also initiates a kind of alienation where 'the body no longer belongs to the person, since ritual is fully a social activity' (112) and in doing so the brahminical philosophy that sustains caste represents, possibly, the most serious denial of the body.

DISTRIBUTION OF AFFECT AND PRODUCTION OF CASTE

Going back to the linkages between space and body, it is imperative for the purposes of this chapter to identify another way in which the body is colonised, which is sensorially (through the senses). Caste-inscribed bodies can also be understood as being regulated under what Joel Lee calls, 'a spatial-sensory order' (Lee 470), where the way in which spaces are organised intensifies the experience of caste for the untouchable body. This phenomenological turn in the study of caste helps us excavate our embodied relationship to our environment or space through sense experience, which in turn further reveals how a space is organised, how the bodies are marked and how a certain ideology, in this case the caste ideology, functions. Therefore, apart from looking at the structural and institutional frameworks that perpetuate caste, we should also take into account aspects of caste that have effects on the moral and aesthetic life constitutive of the full human being. However, before we move on to discuss the spatial–sensory order that corresponds to caste, it might be helpful here to take a theoretical detour and discuss Judith Butler's framework *vis-á-vis* this question of the senses and how affect is regulated through them—a case which proves to be sensitive to our own question.

In *Precarious Life* (2004) and *Frames of War* (2009), Butler develops an account of what she calls a kind of general vulnerability that arises out of a 'fundamental sociality of embodied life', (Butler, *Precarious Life* 28). This is the sociality we are bound to assume as a result of our bodily existence. So, the body according to this view has 'invariably a public dimension constituted as a social phenomenon' (26) and is in perpetuity exposed to others, in addition to always being at risk of violence as a result of that exposure. Given this exposition, Butler argues that injury or violence are thus only the exploitative side of this already existing relationality, or what Aniket Jaaware in his account called the 'bad touch' (*Practicing Caste* 37). Though the exploitative side of this vulnerability is not always on display, the existing vulnerability of the body, Butler admits, becomes highly exacerbated under sociopolitical conditions where violence becomes a way of life (such as the conditions produced from caste inequality).

Therefore, this 'obtrusive alterity' where the 'body invariably comes up against the outside world as a sign of the general predicament of unwilled proximity' (Butler, *Frames of War* 34) animates a responsiveness to that world which may include a range of affects. It is fundamentally in that responsiveness that lies the ultimate responsibility to sustain a moral order. However, when we begin thinking about this responsiveness, we realise that this too is bound up with others. How I respond is essentially linked to who I am, which in turn is related to implicit frames of recognisability. So then, what is our responsibility towards those we don't recognise? This question invites reflection on the exclusionary norms by which fields of recognisability are constituted in the first place, 'fields that are implicitly invoked when by a cultural reflex we mourn for some lives but respond with coldness to the loss of others' (36). For Butler then, 'an ungrievable life is one that cannot be mourned because it has never lived; it has never been counted as a life at all' (38).

This differential distribution of public grieving, she argues, is not just an ethical question but mainly a political issue since grieving itself is bound up with outrage, and outrage often serves as the initial condition on which demands for justice get formulated. Therefore, to understand how some lives are grievable while others are not, 'we need to look at the question of how affect is regulated and of what we mean by the regulation of affect at all' (41). To this, Butler's answer is that our moral responses that take the form of affect are tacitly regulated by certain kinds of interpretive frameworks. Even as it may seem that affective responses are primary and prior to understanding and interpretation, it is actually through existing interpretive schemes that affects are generated and even differentially distributed. How are these interpretive schemes formed? These interpretive schemes that divide life as worthy from unworthy are essentially linked to or fundamentally operate through the senses at the level of touch, smell, sight and so on, where all of our senses are regulated in different ways under a given interpretive framework. Now *vis-á-vis* the interpretive framework of caste we can take the example of how smell and space play a very important role in the way that this differential distribution of affect via the senses is regulated.

Joel Lee (2017) and others like Hugo Gorringe and Irene Rafanell (2007) have conducted ethnographic work on caste especially regarding questions of filth, odour and public spaces relevant to our present discussion. Joel Lee argues that 'smell' cannot be simply taken as a rhetorical trope at the level of discursive knowledge where one is, say, taught that certain people by nature smell a certain kind of way, because that would ignore the real domain of smell and the phenomenological experience behind it. Therefore, while looking at the empirical domain of smell, one can observe in a caste society how smell is organised in a way that reproduces caste order through a sensory experience. This structures the everyday experience of the untouchable body, inculcating habits of actions that correspond to their assigned social location forming what Bourdieu might call a 'caste habitus' (Gorringe and Rafanell 103). Caste hierarchy through smell is given an embedded materiality through which caste becomes more visible and, in that sense, more real.

To see how caste translates into a spatial–sensory order, one can look at both the social morphology of the village and the city where, through what Lefebvre calls certain 'spatial practices' a concrete sensorial order is produced. For instance, one can see distinct residential patterns in the village where the 'dalit dwellings' are segregated from the main village and located on the margins. This is both a space apart and a space to pollute where all the waste is either dumped or public toilets are located, or it is a site demarcated for cremation which also depends on the untouchable's labour for its management. So, the untouchable's place of work is made to coincide with their place of residence. Most of these practices produce odour and by the virtue of these odorants being regularly dumped and relatively stabilised in the same location, smell assumes a spatial quality. Villages almost produce an olfactory map where the olfactory and spatial divisions in the village are made to overlap with social divisions.

Similar residential patterns are reproduced in the city with regards to sanitation workers, where the infrastructure of waste overlaps with the housing provided for these workers. So even with the coming of modernity and increased migration leading to a chance for former untouchables to move to urban spaces, they were upon arrival at once included through jobs and excluded through housing, where a residential pattern similar to the village was reproduced.

Thus, the way that affect impinges on the spatial dialectic of caste-inscribed bodies is via the senses. As a spatial–sensory order is established through the interpretive scheme of caste, space gets merged with various sensory connotations, as shown in the case of the space of a village with a dalit population. This inaugurates an order of affect whereby, as Butler argues, some lives become grievable while others are left out. In lieu of a conclusion, we can maintain that the reconfiguration of space as discussed above continues to remind us of the various ways in which caste endures even in the modern world and is not, as often suggested, a relic of the past which the history of modernity has liquidated, or a feature of only Indian villages. Space is a reminder of caste, as it makes visible what is invisible and unconceals what is concealed in finite, enclosed and divided sites. Even in urban spaces, the dalit body finds no solace in busy streets and is almost always haunted by the ghost of caste that follows them like a shadow through their sensorial perceptions.

As space becomes a means through which caste continues to sustain itself, it nonetheless provides the scene for much of one's action and thought. The absence of space that allows for free relations to take shape, be it in the villages or the city, ultimately results in stunting bodily growth which might manifest itself through diseases that are borne out of proximity to actual filth. But, most importantly, experience of space through caste impedes the life of the mind which is ultimately a life of dignity and self-respect. It is only by re-thinking space, as Arendt suggests, that we can actualise our human capacities as creators and makers of the world.

Works Cited

Ambedkar, B.R. "The Untouchables: Who were they and why they became Untouchables?". *Dr Babasaheb Ambedkar: Writings and Speeches*, edited by Vasant Moon, vol. 7, Govt. of Maharashtra, 1990, pp. 239–380.

———. "Untouchables or the Children of India's Ghetto". *Dr. Babasaheb Ambedkar: Writings and Speeches*, edited by Vasant Moon, vol. 5, Govt. of Maharashtra, 1989, pp. 3–126.

———. "Waiting for a Visa". *Dr. Babasaheb Ambedkar: Writings and Speeches*, edited by Vasant Moon, vol. 12, Govt. of Maharashtra, 1993, pp. 661–692.

———. "Who were the Shudras?". *Dr. Babasaheb Ambedkar: Writings and Speeches*, edited by Vasant Moon, vol. 7, Govt. of Maharashtra, 1990, pp. 5–209.

Arendt, Hannah. *The Human Condition*. U of Chicago P, 1998.

Banerjee, Prathama. "Caste and History: Writing in India". *Dalit Assertion in Society, Literature and History*, edited by Imtiaz Ahmed and Shashi Bhushan Upadhyay, Deshkal Society, 2010.

Butler, Judith. *Frames of War: When is Life Grievable?*. Verso, 2009.

———. *Precarious Life: The Powers of Mourning and Violence*. Verso, 2004.

Cokal, Susann. "Wounds, Ruptures and Sudden Space in the Fiction of Georges Bataille". *French Forum*, vol. 25, no. 1, 2000, pp. 75–96.

Douglas, Mary. *Natural Symbols: Explorations in Cosmology*. Routledge Classics, 2003.

Duneier, Mitchell. *Ghetto: The Invention of a Place, The History of an Idea*. Farrar, Straus and Giroux, 2016.

Foucault, Michel. *Power: Essential Works of Foucault 1954-1984*. Translated by Robert Hurley, edited by James D. Faubion, vol. 3, New Press, 1997.

Gorringe, Hugo, and Irene Rafanell. "The Embodiment of Caste: Oppression, Protest and Change". *Sage Journals*, vol. 41, no. 1, 2007, pp. 97–114.

Guru, Gopal. "Archaeology of Untouchability". *Economic and Political Weekly*, vol. 44, no. 37, 2009, pp. 49–56.

———. "Experience, Space, and Justice". *The Cracked Mirror: An Indian Debate on Experience and Theory*. Oxford UP, 2012, pp. 72–104.

Jaaware, Aniket. "Eating, and Eating with, the Dalit: A Reconsideration Touching on Marathi Poetry". *Indian Poetry: Modernism and After*, edited by K. Satchidanandan, Sahitya Akademi, 2001, pp. 262–93.

———. *Practicing Caste: On Touching and Not Touching*. Fordham UP, 2019.

Kumar, Aishwary. *Radical Equality: Ambedkar, Gandhi, and the Risk of Democracy*. Stanford UP, 2015.

Lee, Joel. "Odor and Order: How Caste is Inscribed in Space and Sensoria". *Comparative Studies of South Asia, Africa and Middle East*, vol. 37, no. 3, 2017, pp. 470–90.

Lefebvre, Henri. *The Production of Space*. Blackwell, 1991.

Manjali, Franson. "The Body of Sense, the Sense of Body". *Rivista Italiana di Filosofia del Linguaggio*, vol. 2, 2010, pp. 95–122.

Ramanujam, Srinivasa. *Renunciation and Untouchability in India: The Notional and the Empirical in the Caste Order*. Routledge, 2020.

Sarukkai, Sundar. "Phenomenology of Untouchability". *The Cracked Mirror: An Indian Debate on Experience and Theory*. Oxford UP, 2012, pp. 157–99.

Vinayaraj, Y.T. *Dalit Theology After Continental Philosophy*. Palgrave Macmillan, 2016.

FIFTEEN

A SENSE OF WASTE

ENVIRONMENT AS ALIENATED 'SETTING' AND BENGALI EXPERIMENTAL FICTION

Samrat Sengupta

Introducing the Man/Environment Difference in Narratives

The secret premise of modernity is the alienation of man from his environment. The alienation is an outcome of the crisis produced from assumed mastery and control over nature by the he-man, the imagined phallocentric human self. Humans are technological creatures constantly producing and extending the meaning of their being and the world or being-in-the-world (in conjunction). They are creatures of de-fault—the original fault of not having a specific 'naturally' given role on earth (Stiegler 16–17). They have to define their role and also reconstruct the world in terms of their being in it. In this sense, ontology and epistemology share an intricate relationship with narration, with storytelling that connects the humans with the world, defining it both in terms of relationship and difference. Phenomenological descriptions of the world relate humans with their surroundings and reconsider knowledge as relational. It authenticates knowledge in terms of an original connection and situatedness in the world. However, these descriptions have to share a deep relationship with language and narration in order for that connection to be established (Wylie 47). Man talks about nature scientifically or poetically, and in each discourse nature is an object from which man, even if connected, remains individuated as a separate being. But if man is also a part of the natural ecology, then his separation waits to be reconfigured in terms of his situatedness to the space of his belonging. This produces a failure of narratives and of absolute knowledge of nature. Man realises his poverty with respect to both his being and the world in which he claims to dwell.

Can we then have an eco-phenomenology (Kerridge 66–68; Wood 211–33) which will make our narratives more precarious and unsure?

How do we look at the relationship of human stories with the settings of a narrative—the natural and naturalised background (both natural and social ecology)—in such a precarious form of a relationship? 'Nature' as a concept distances man from the environment, of which he is a part. We may however challenge this assumption and see how nature is also produced in the process of world making and the self-making of man (Morton, *Ecology Without Nature* 1–28; *The Ecological Thought* 1–19). This may also orient our idea of space as secondary to human life existing only to serve their purpose or defined and designed by man to being suited to human use. This making suitable is done by appropriating space as natural or without nature (for example, in urban spaces where the presence of natural objects, like plants or trees, that often form a part of a natural landscape is minimised). In both cases, space is defined in terms of human use, the desire and longing for a particular kind of space and shaping it in that way. We can argue that such a conceptualisation is manufactured and maintained through certain kinds of narrativisation. The objective of this chapter is to see if counter-modernist practices may develop outside Western epistemological responses to an industrial Capitalocene-inducing modernity. Experimental storytelling could be seen as an attempt to break the hierarchy between man and environment or the human and the natural world, challenging the humanist framework of the primacy of man in narratives. Some of the works of Subimal Mishra, a late-twentieth century Bengali experimental writer, will be employed here to illustrate such possible counter-intuitive practices of narration. We will see how the limits of fictional storytelling are redrawn and how the difference between an anthropocentric plot and its respective setting gets unsettled in Subimal's works.

THE ANTHROPOCENTRISM OF STORYTELLING AND WORLDING

Whether it is fictional or non-fictional, whether it is subjective (in the form of autobiographies or testimonies) or objective (like history or sociology), stories mostly divide the human world of action and the world around them which provides the background or setting to such action. The human becomes a part of that world only by this separation. This self-making is also world-making, or what Heidegger calls 'worlding' (143–212). By alienating the conceptually uninscribed 'earth', the 'world' is born. This world is born as human writing—

as spaces translated into places by the human cognitive apparatus. Writing is also a process of unwriting what it alienates and negates in the process of signification. That alienated other of writing works as a trace of the narrative. We may argue here that the 'setting' in a story is such an 'other' of the human narratives it is internal and intrinsic to. It is a process of place-ing the story—rooting the subject of the story, reducing the environment (both humanly produced and natural) into an object of use. The human subject stands erect in such a 'background'. The background is often nature, but it can also be unnatural objects, 'hyperobjects' produced by man over which he has no control (Morton), or it could be other humans reduced to the status of nonhuman or to objects. Nature is touched by man both ideologically—by creating the binary of natural and social—and practically by shaping nature either through indiscriminate developmental activities or through protectionist intervention. In both cases nature becomes an object of human calculation, speculation and judgement. The dominant structure of most narratives naturalises both nature and society as spaces where the human subject dwells. They orchestrate this by producing the setting of a narrative as the naturalised background where human characters are placed and made to act.

The contention here is to show if and how stories can constantly move towards overcoming such alienation between man and environment thereby challenging the humanist framework of thinking and moving towards a posthuman framework. The human/natural difference becomes unsettled in such thought, and it may help to restructure the value system of narratives. In this chapter, we try to show how the difference between event and setting, or between human actors and their nonhuman background, can be problematised through an active restructuring of the narrative—experimental in form and political in approach. The politics here lies in challenging the alienation of environment as an essential requirement for stories to exist. Experimental writing is perhaps one particular form and method that could be used to challenge such value inscription in narratives. Wendy Knepper and Sharae Deckard comment that 'experimental writing self-consciously locates its aesthetic interventions within an unevenly constituted global literary marketplace for experimentation in which some literary conventions and knowledges prevail over others, such that aesthetic choices are necessarily politicized' (4). It contains the potentiality to question the existing structure of alienation, not only of man but of all 'disposable beings'. In this formulation, instead of being a special victim of alienation by other humans—a state that needs to be politicised and contested thereby foregrounding the ethics of literary writing—man becomes a part of the generality of disposable beings, wasted by the human economy of use. Therefore, we need to understand the disposability of humans, nonhumans and other beings alike within the structures of enframed human imagination and understand how this disposability is sanctioned through regular modes of narration. The attempt made in this essay is to show how experimental writing may unsettle such assumptions by untying the elemental connections between characters, events and settings within a text.

Writing about environment is also often inscribed in an ego–logical fallacy, where man laments the damage done to nature by his species and hopes for a reconciliation, or at least tries to negotiate with that damage. Romanticism and modernism are two ends of an Anglo-European response to industrial modernity and the Capitalocene, where global capital touches every nook and corner of earth and transforms it irredeemably. Heather Davis and Etienne Turpin write, 'The Capitalocene ... points directly to a voracious political economic system that knows no bounds, one where human lives, the lives of other creatures, and the beauty and wealth of the earth itself are figured as mere resources and externalities' (7). If Romanticism promises escape to a lost 'nature', modernism laments an incommensurable and incurable loss of it thereby sharing a negative relationship with environment. Both use 'nature' as a ploy to establish a relationship of difference between man and their surroundings thereby keeping man at the centre of the axis of interpretation. However, the modernity induced by the Capitalocene might not allow for such easy separation since, while nature within human ecology is decrepit and damaged, nature outside is already contaminated by traces of human activities.

To overcome this economy of use, to challenge this ego–logical hierarchy of man and the environment—where environment is a subtext to the human story or a foil—two strategies could be imagined, one in terms of content and narration/use of language and another in terms of form and structure. In this chapter I choose examples from the world of modern Bengali literature which may claim to create an alternative version of modernity, challenging dominant forms of literary writing. In the process, such literary works actually produce political responses to the 'combined and uneven development of world' under global capitalism (Warwick Research Collective) in their experimentation with form. These writings entail a philosophical commentary on forms of alienation sustained by world capitalism, maintained in dominant styles of writing. The separation of man from the environment is one such stylistic hegemony in literature that creates the essence of naturalist poetics. Inside such a poetics, man is either naturally an uninterrupted part of nature or suffers alienation from their environment by removal from that nature. It does not however question the human domination of earth both by occupying lands and by colonising it through naming and meaning making. Ego–logical storytelling is yet another colonisation of environment as spaces are reduced to a human purpose of telling stories about themselves. Experimental writing can dismantle that process of storytelling. One way of doing this is through a non-hierarchical placing of human and nonhuman worlds or a reversal of the hierarchy between them, initiating a posthumanist aesthetic form in narration. This reversal of hierarchy between the human and the nonhuman world may happen in two possible ways: with the intrusion of the human into nature making the story of nature always and already contaminated; and the contamination of the human world by nature. This 'nature' has already succumbed to the sacrilege of irreducible human trace and at the same time continues to affect the human world. So separation is impossible, though alienation becomes the architectonic of the man–nature relationship.

ALIENATED SELF AND THE LOSS OF 'NATURE'

A short story 'Pakhi' ('Bird', 1971) by the Bengali experimental writer Subimal Mishra depicts the ecological alienation caused by man to nature. Here the human–environment binary is sustained, but it employs a counter-aesthetic practice where the human subject and their desire for prelapsarian connection with nature gets unsettled. The story is about a boy, Bhutu (in English the word suggests a spectre) who sets out to see birds in flight along with his companions. He follows and walks along a railway track in search of birds but cannot find one. In the end, in a place occupied by ruins, as the day ends and darkness falls, he finds a dead bird at his feet. It has been shot by a bullet. But instead of steel the bullet is made of burnt clay. Bhutu remembers that in his childhood he used to hunt birds with pieces of brick shot from a catapult. Here we witness that instead of an ego–logical narrative, the subject confronts a primordial human crime, more ancient than has been imagined. The modernist understanding of a postlapsarian time is also challenged. The human crime of alienation goes beyond chronological imagination. By unsettling the chronology of the modern subject, his relationship of separation from his environment is punctured. It is man who creates their environment in a specific way. Ranjan Ghosh in his article 'Plastic Literature' discusses a retelling of Coleridge's "Rime of an Ancient Mariner" through an apocalyptic postreading. It can be thought of as postreading since the primacy of the human subject is decentred from the process of reading, making it an impossible adventure.

In Nick Hayek's 'Rime of a Modern Mariner' the dead Albatross is 'shot by a different kind of human agency: man's discovery of a substance about which he does not know what to do' (Ghosh 283) and its stomach contains plastic waste from a wrecked ship almost eighty years old. Eventually, the movement of the ship gets clogged by plastic waste. There is no forgiveness for this human crime and no trace of 'blue sky bending over all', showering divine grace in the end the way we see in Coleridge's poem. Plastic waste appears as a ghost. It unsettles our imaginary separation and linear relationship between past, present and future, as the past of wasted things haunts the present of the 'human' in an unanticipated way and shadows the future as an uncanny 'writing back' of human history by the alienated environment. Ghosh comments, 'If plastic was the surprise of the *anthropos* at the beginning of the last century, plastic today is the surprise that has met it from the other side. The counterinsurgency is the uncanny dehumanizing process of writing back the history of planet life, the planet discourse ...' (284). The bird becomes the bearer of that uncanny spectre of plastic—the human crime as well as the future. The separation of the human from the environment is jeopardised as the hierarchical dominion of the human on nature and his alienation from it is written back. When the human's crime on nature returns to him, consuming the albatross and the human alike, the hierarchy collapses.

In Mishra's story the sinister and pathological relationship of man with his environment is detemporalised as it is sketched out of the project of modernity. Bhutu, the human protagonist, finds a dead bird shot by his catapult in childhood returning to him in a birdless world. He finds it when he has stopped his lookout for birds in flight and is searching for his long-lost doll. The desire for a secure relationship with a doll-like object or long-lost nature and the romanticism of witnessing a bird's flight is interrupted by the primordial human crime of reducing environment to a human object of use. The spectre of that human crime distances Bhutu forever from his long-lost village which he can only enter by following the path of the railway tracks through the preordained chequered formulae of a modern bird watcher.

In modernist Bengali film maker Satyajit Ray's film *Pratidwandi* (1972, based on a novel of the same name by Sunil Gangopadhyay, a popular Bengali fiction writer) the modernist hero Siddhartha is in search of a bird whose whistle he heard in his childhood. He goes to Esplanade, where a famous marketplace of Kolkata is located, looking for this bird. The scene starts with broiler chickens being thrown and packed in a basket and then we have a series of cages with variety of birds on display. Siddhartha is disgusted as he cannot identify the bird. The film ends with the scene of Siddhartha leaving the city and starting work as a village postmaster. There he rediscovers that bird.

There is a prelapsarian turn in Siddhartha's life amidst nature. This does not happen in Subimal Mishra's 'Pakhi'. Bhutu, as his name suggests, is a spectre of himself—a spectre born of the process of being human. Unlike in the modernist framework, he is not a victim of alienation but also a maker of it. The primordial human crime returns to him with a certain sense of the uncanny—'Trembling in excitement, Bhutu realizes that the dead bird has been hurt not by a steel bullet, but a bullet made of burnt clay' (163; translation mine). The solace of a pristine Heideggerian 'earth' poised against the worlding of the world is taken away, as the earth also is inscribed and written by the human economy of use and becomes a weapon in the form of 'burnt clay'. Human use itself creates a primordial drift that naturally assumes nature to be an object of use.

HOW TO WRITE THE WASTED BEINGS?

So far, we have seen that fracturing human temporalities and bringing back the effects produced by man on nature to them in an uncanny spectral way is a form of questioning the neatly divided human–environment or subject–object binary. Another way can be breaking the human–nonhuman binary by foregrounding the story of wasted beings; by placing the wasted and alienated environment—animals, objects, humans—together in a synchronic and non-hierarchical state of belonging within a narrative. The crisis of writing becomes evident as these 'non-beings' interrupt the process of storytelling, so far dominated by the human subject. The background becomes the story; the setting itself becomes the narrative.

Subimal's stories have several such collapses where wasted humans are placed beside damaged environments. A cadaver appears in many of his stories. The dead body becomes an apt image of the expendability of man in the economy of use. Man is just another potentially wasteful object in this order, on the side of the environment, continually succumbing to the cycle of use and abuse. In Subimal, the damaged beings (either human or nonhuman) are often chosen from the lowest rung of the society—beggars, pavement-dwellers, refugees, people, living in shanties and those devastated by economic and social exploitation. They become one with the environment in terms of damage and decimation, decay and wastefulness. They could be read as a warning to the human world order, with a promise of apocalypse or revolutionary upturn.

In Subimal's short stories such as 'Rakter swabhab' or 'The nature of blood' and 'Nanga haar jege uthche' or 'Naked bones awakening', we see the homeless from different parts of Bengal and refugees from East Pakistan eroding the city and occupying it. In 'Nanga haar jege uthche' the inroad of homeless people is paralleled in a montage with a possible deluge as water crosses the threshold of the Ganges and starts overflowing the city. The apocalyptic vision of the unwanted people crossing the limits of state and governmentality enmeshes and parallels the unanticipatibility of nature—a water level that may cross its threshold any moment.

David Arnold in *Toxic Histories,* in the context of nineteenth century Indian cities, discusses a certain '"bourgeois environmentalism", in which urban pollution and environmental degradation are seen as coming, in no small measure, from the occupations and habitations of the poor' (191). However, poor people are the worst victims of pollution and toxicity. Their relationship with the environment is one of damage and decimation. Arnold further writes about post-industrial Indian cities of the nineteenth century: 'In reality, those who were most at risk from the pollution and poison were often the workers themselves, the poor, low-caste men, women and children who toiled in lime kilns, tanneries and dyeing works with little, if any, protection against contact with caustic substances or the inhalation of toxic fumes' (190).

The situation has changed only a little in contemporary India. Scholars like Joel Lee or Mukul Sharma have raised the question of environmental casteism or eco-casteism while discussing modern day India. Bourgeois environmentalism wants to keep toxic environment and people out of the sphere of bourgeois security. The association of the threat of unwanted people and other beings with impending environmental catastrophes, like that of a flood, in literary writing collapses the human–environment difference and pushes the entire compendium of existence towards an unanticipatable precarity. The image of a dead cadaver functions as a threat since dead bodies are often understood as a source of toxicity and pollution. They also break down the ego–logical and anthropocentric discourse. Like the expendable homeless poor and 'lower-caste' population who are more a part of the polluted environment than an affirmation of human subjectivity, the cadaver signifies the dead 'human' who becomes a part of the polluted environment—a visible threat to the rights-bearing bourgeois human subject. Janam Mukherjee writes on Subimal Mishra, 'his writings continued to push at the boundaries of narrative device, and also the boundaries of middle-class, bourgeois mores, structured towards the maintenance of a pernicious and predatory socio-political order that ensures both the destruction and erasure of lives lived at the margins of bare existence' (Mukherjee ix). However, one may argue that the margins in Subimal's work do not include living or sentient beings only but also the environment they live in. He collapses the ideological hierarchy between life and non-life in his method of narration.

FRAGMENTED NARRATIVES: THE CONTAMINATED WORLD AND SELVES

Beside the literary relationship of human and environment being laid in terms of difference and alienation—or the archetype of alienated, objectified and damaged nature casting its shadow on 'other' humans who are marginalised and exploited in the existing sociopolitical order (thereby unsettling the hierarchy of human–environment with marginal humans being identified with a damaged and decimated environment)—there can be a third imaginary relationship. This is a relationship of narrative fragmentation where, not only in terms of content but also in terms of the narrative structure, environment becomes as much a subject as the human. Subimal Mishra, in one of his essays, delineates this new relationship between multiple elements in fiction, the human subject being one of them. In 'Uponyas, anti-uponyas, uponyaser mrityu' ('Novel, anti-novel, death of the novel) he comments,

> Since the human subject collects elements of contradiction from society and explodes like a seed, individual pain becomes social. Everything mixes up into a mould. The country fuses with the city, superstition with

> scientific mindscape, subjectivity with matter, the real with the surreal, everything comes together in a montage ... Then this infusion becomes a punch whose keyword is contradiction (391–92; translation mine).

Mishra's narratives are such medley of beings and things—such a concatenation of people and environment that the reader is often at a loss. His stories make environment a part of the narrative or they put the idea of narrative at risk. Here two such narrative techniques should be mentioned. One is his lack of punctuation, which resists a complete reading of the shifts or cuts between different narrations of the story that may include a supposed main plot, imaginations and nightmares of the characters, and their various spaces of dwelling which do not simply act always as backgrounds.

In his novella, 'Tejoshkriyo aborjona' or 'Radioactive waste', Mishra brings several such narratives, discourses and descriptions together. This text was written in 1970 and published in 1972 with the title 'Technical manifesto'. Until earth-shattering nuclear power catastrophes such as Chernobyl or Fukushima happened (where nuclear power was meant for domestic use and not war), the discourse of nuclear fear was confined mainly to the threat of explosion and war (Yoshimoto; Nancy; Cordle). Jean-Luc Nancy writes, 'The problem posed by the "peaceful" use of the atom is that of its extreme, and extremely lasting, harmfulness' (18). Much before literature opened up the space of the everyday to nuclear fear, Mishra authored this story where radioactive contamination is made a part of everyday human discourse which possibly 'suggest(s) an enmeshment of human and natural world that is tied up ... with something going awry' (Cordle 291).

Among the fragments of various linear narratives appearing randomly and moving in parallel in Subimal's story, one is that of an affair between a young man Ajay and a slightly elderly lady Sushoma. They stay together in Ajay's room secretly when Ajay's elder brother and sister-in-law are not at home. Their space of intimacy is disturbed and interspersed by a ruckus outside as some rival gangs throw bombs at each other. We also get to know during the course of the narrative that Ajay is not fond of the 'fat woman Sushoma' who is pregnant with his child. The story is discontinuous and penetrated with other narratives which include a description of human cell biology; how a zygote is formed inside Sushoma's womb; mythological tales; stories of violence committed by US soldiers in the Vietnam War; and finally the description of radioactive pollution. Lack of punctuation makes the discontinuous narration seamless and oddly threaded into one world. The human damage to the environment through the use of nuclear power plants contaminates everything, including the human body. No room of one's own is secured from ecological as well as political pollution. Human and nonhuman worlds alike have to bear the consequence of man's thirst for power, greed, lust, violence and pride. It suggests an '"interdependence" between human and natural worlds that challenges our tendency to think of them as separate things' (Cordle 291). Subimal comments in the story that while advanced cell biology tries to make artificial humans, nuclear waste is threatening life on earth as a whole.

Curiously, a portion of the last section of the story describing the effects of nuclear waste on human and natural world is incorporated as the image of a tattered old page before the beginning of the story (Figure 15.1). The title 'Radioactive waste' in Bangla font and the name of the author are underlined and marked with a tick on that tattered page. This possibly suggests the need for punctuating the human story with a larger discourse of environmental damage, impossible to be contained in a narrative. The placing of the tattered page before the story perhaps indicates the contamination and wasting of nature as an ancient and atemporal crime that comes before the beginning of the human story. Its origin is disappearing like a tattered old page, though it comes back to the future. The Anthropocene and Capitalocene are not matters of past or present and contaminate the future to come. It perhaps suggests that like unclaimed and uncertain plastic waste, nuclear waste also casts a shadow on the human story and makes it a non-story by linking it to the larger environment: damaged and decayed by human trace, no longer traceable. It is an atemporal fragment that resists the formation of a linear narrative and contaminates stories as much as they transubstantiate the human body.

The final page of the novel contains a table of radioactive substances and how it flows from the atmosphere through plants and animals into human body. It acts as a sinister warning against the imagined separation of man and his environment. This pushes us towards a posthuman hermeneutics of the human–environment relationship where a continuity between the human and nonhuman world has been presumed. The damage done to environment, which bears the traces of man-made hyperobjects, may precariously return to the human body unknowingly and inscrutably. Contaminated nature enters into a homological relationship with the human body, as the

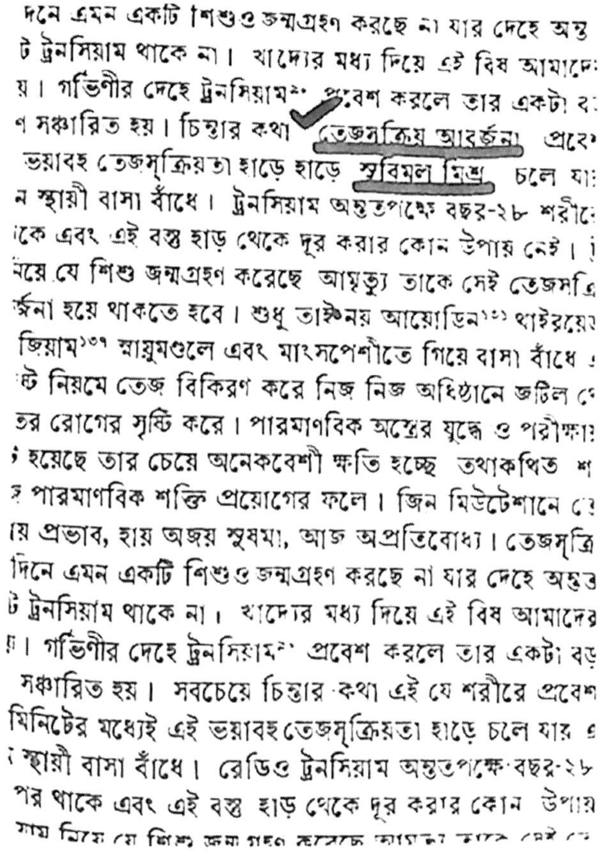

দনে এমন একটি শিশুও জন্মগ্রহণ করছে না যার দেহে অন্ত
ট ট্রনসিয়াম থাকে না। খাদ্যের মধ্য দিয়ে এই বিষ আমাদে
য়। গর্ভিণীর দেহে ট্রনসিয়াম৯০ প্রবেশ করলে তার একটা ব
ণ সঞ্চারিত হয়। চিন্তার কথা তেজস্ক্রিয় আবর্জনা প্রবে
ভয়াবহ তেজস্ক্রিয়তা হাড়ে হাড়ে সুবিমল মিশ্র চলে যা
ন স্থায়ী বাসা বাঁধে। ট্রনসিয়াম অন্ততপক্ষে বছর-২৮ শরীরে
কে এবং এই বস্তু হাড় থেকে দূর করার কোন উপায় নেই।
নয়ে যে শিশু জন্মগ্রহণ করেছে আমৃত্যু তাকে সেই তেজস্ক্রি
র্জনা হয়ে থাকতে হবে। শুধু তাই নয় আয়োডিন১৩১ থাইরয়ে
জিয়াম১৩৭ স্নায়ুমণ্ডলে এবং মাংসপেশীতে গিয়ে বাসা বাঁধে
ষ্ট নিয়মে তেজ বিকিরণ করে নিজ নিজ অধিষ্ঠানে জটিল
তর রোগের সৃষ্টি করে। পারমাণবিক অস্ত্রের যুদ্ধে ও পরীক্ষা
; হয়েছে তার চেয়ে অনেকবেশী ক্ষতি হচ্ছে তথাকথিত শ
র পারমাণবিক শক্তি প্রয়োগের ফলে। জিন মিউটেশানে
য় প্রভাব, হায় অজয় সুষমা, আজ অপ্রতিবোধ্য। তেজস্ক্রি
দিনে এমন একটি শিশুও জন্মগ্রহণ করছে না যার দেহে অন্তত
ট ট্রনসিয়াম থাকে না। খাদ্যের মধ্য দিয়ে এই বিষ আমাদের
। গর্ভিণীর দেহে ট্রনসিয়াম৯০ প্রবেশ করলে তার একটা বড়
সঞ্চারিত হয়। সবচেয়ে চিন্তার কথা এই যে শরীরে প্রবেশ
মিনিটের মধ্যেই এই ভয়াবহ তেজস্ক্রিয়তা হাড়ে চলে যায়
স্থায়ী বাসা বাঁধে। রেডিও ট্রনসিয়াম অন্ততপক্ষে বছর-২৮
পর থাকে এবং এই বস্তু হাড় থেকে দূর করার কোন উপায়

Figure 15.1: The image of a tattered page placed before the novella 'Radioactive waste'. (Mishra, *Boi Sangraha-2* 343)

setting and plot become indistinguishable in Mishra's fiction, thereby resisting a careful charting out of human and natural world separately. The inner and outer pollution run in parallel and penetrate each other. The ideology of disposability causes the man in the story to find ways of avoiding his pregnant partner while the environment enters the human story in terms of radioactive waste assigning the mark of human abuse to nature.

Discussing the first and second waves of ecocriticism, Robert Brazeau and Derek Gladwin observe that while the first wave 'focus(es) on literary texts that address both general and specific issues dealing with the environment, language and culture' (3), the second wave 'offered new ideas about how this theory can speak to issues beyond the aesthetic and literary. Ecocritical works frequently engage with ethical, practical and political concerns' (3). Quoting Lawrence Bruell, Brazeau and Gladwin combine the two and talk about an 'environmental criticism' where 'a mature environmental aesthetics—or ethic, or politics—must take into account the interpenetration of metropolis and outback, of anthropocentric as well as biocentric concerns.' (qtd. in Brazeau and Gladwin 4). While Bruell talks about the politics of writing about environment, its ethics, aesthetics and politics, here following Mishra's use of narrative dissonance and montage based on inner contradiction, we may think, instead, of an eco-narratology where natural and moral ecology cannot be separated from the human. Instead, the process of worlding employs a seamless contamination of earth, man and other beings and objects by each other. The damage cannot be traced as a linear narrative with a clearly known human beginning. Rather it shares a spectral relationship with all stories where humans and nonhumans have a shared fate of decay and contamination. This decay and contamination is situated, not only in the content, but also in the form where each micronarrative pollutes the other. As suggested by the novella 'Radioactive waste', it is governed by the haunting of a spectre—the spectre of toxicity and waste. The waste eventually cannot be separated from human narratives. Humans become a part of narrative waste.

Works Cited

Arnold, David. *Toxic Histories: Poison and Pollution in Modern India*. Cambridge UP, 2016.

Brazeau, Robert and Derek Gladwin. Introduction: James Joyce and Ecocriticism. *Eco-Joyce: The Environmental Imagination of James Joyce*, edited by Robert Brazeau and Derek Gladwin, Cork UP, 2014, pp. 1–17.

Cordle, Daniel. "Climate Criticism and Nuclear Criticism". *Climate and Literature*, edited by Adeline Johns-Putra, Cambridge UP, 2019, pp. 281–97.

Davis, Heather and Etienne Turpin. "Art & Death: Lives Between the Fifth Assessment & the Sixth Extinction". *Art in the Anthropocene: Encounters Among Aesthetics, Politics, Environments and Epistemologies*, edited by Heather Davis and Etienne Turpin, Open UP, 2015, pp. 3–29.

Ghosh, Ranjan. "Plastic Literature". *University of Toronto Quarterly*, vol. 88, no. 2, 2019, pp. 277–91.

Heidegger, Martin. "The Origin of the Work of Art". *Martin Heidegger: Basic Writings*, edited by David Farell Krell, Harper Perennial, 2008, pp. 243–63.

Kerridge, Richard. "Environmental Fiction and Narrative Openness". *Process: Landscape and Text*, edited by Catherine Brace and Adeline Johns-Putra, Rodopi, 2010, pp. 65–85.

Knepper, Wendy and Sharae Deckard. "Towards a Radical World Literature: Experimental Writing in a Globalizing World". *Ariel: A Review of International English Literature*, vol. 47, nos. 1–2, 2016, pp. 1–25.

Lee, Joel. "Odor and Order: How Caste Is Inscribed in Space and Sensoria". *Comparative Studies of South Asia, Africa and the Middle East*, vol. 37, no. 3, 2017, pp. 470–90.

Mishra, Subimal. "Nanga haar jege uthche (Naked bones awakening)". *Boi Sangraha-1*. Gangchil, 2012, pp. 95–101.

———. "Pakhi (Bird)". *Boi Sangraha-1*. Gangchil, 2012, pp. 160–63.

———. "Rakter swabhab (Nature of blood)". *Boi Sangraha-1*. Gangchil, 2012, pp. 72–81.

———. "Tejoshkriyo aborjona (Radioactive waste)". *Boi Sangraha-2*. Gangchil, 2013, pp. 341–66.

———. "Uponyas, anti-uponyas, uponyaser mrityu (Novel, anti-novel, death of the novel)". *Boi Sangraha-2*. Gangchil, 2013, pp. 391–401.

Morton, Timothy. *Ecology Without Nature: Rethinking Environmental Aesthetics*. Harvard UP, 2007.

———. *The Ecological Thought*. Harvard UP, 2010.

———. *Hyperobjects: Philosophy and Ecology After the End of the World*. U of Minnesota P, 2013.

Mukherjee, Janam. Introduction. *Actually This Could Have Become Ramayan-Chamar's Tale*, by Subimal Mishra, translated by V. Ramaswamy, Harper Perennial, 2019, pp. vii–xvi.

Nancy, Jean-Luc. *After Fukushima: The Equivalence of Catastrophes*. Fordham UP, 2015.

Ray, Satyajit, dir. *Pratidwandi*. Priya Films, distributed by Shri Bishnu Pictures Private Limited, 1970.

Sharma, Mukul. *Caste and Nature: Dalits and Indian Environmental Politics*. Oxford UP, 2017.

Stiegler, Bernard. *Technics and Time, 1: The Fault of Epimatheus*. Stanford UP, 1998.

Warwick Research Collective. *Combined and Uneven Development: Towards a New Theory of World Literature.* Liverpool UP, 2015.

Wood, David. "What is Eco-Phenomenology?". *Eco-Phenomenology: Back to the Earth Itself*, edited by Charles S. Brown and Ted Toadvine, State U of New York P, 2003, pp. 211–33.

Wylie, John. "Writing Through Landscape". *Process: Landscape and Text*, edited by Catherine Brace and Adeline Johns-Putra, Rodopi, 2010, pp. 45–64.

Yoshimoto, Mitsuhiro. "Nuclear Disasters and Invisible Spectacles". *Asian Cinema*, vol. 30, no. 2, 2019, pp. 169–85.

GLOSSARY

akhada: the gathering place in a Santal village where people meet; this space is also typically used for dancing.

anthropomorphism: the act of projecting human-like qualities on other-than-humans, thereby integrating those entities into humans' own sensory lives; as such projections often lack codified reasoning or evidence, they are usually rejected by scientific theories and methods.

Asian Elephant: the only elephant species distributed across South and Southeast Asia; *Elephas maximus* is morphologically and genetically distinct from the forest/bush and savannah elephants of Africa. Once abundant across the Indian subcontinent, the present-day Asian Elephant populations are severely fragmented due to large-scale habitat loss across its distribution range. Northeast India harbours close to one-third of the total Asian elephant population of India.

attunement: to be attuned to the other is to make oneself properly aware or responsive so that one fits into an arrangement of harmonious correspondence with them; attunement is a priming of the embodied self to be available, open and free—stretched taut like a string—to the touch of the other.

birdwatching: Birdwatching is a hobby and citizen sport that involves observing live birds in their natural habitat. It involves identification of a bird's species, description of their plumage and parts, nesting habits and social behaviour. Birders carry a field guide to help them in identification of birds, a notebook to record their own observations and a pair of binoculars.

Capitalocene: Jason W. Moore's idea of the Capitalocene supplements the theory of the Anthropocene as a new geological era induced by traces of human activities damaging the environment of the earth permanently. The Capitalocene claims that such decimation is orchestrated by capitalism and its ideology is to assume everything on earth as meant for human use or consumption, and to further the accumulation of capital.

colonial modernity: modernity, as understood to be a project of the West and the institutions and bases of which are deeply rooted in colonial practices and politics, which impresses itself on the colonised world; it also engenders a temporal notion of progress marked by economic and political registers of the colonial project.

cordon sanitaire: a quarantined space that restricts the movement of people from going in and out of that specific area; this phrase was instructive for Ambedkar to think about the institution of caste in so far as it prohibited movement and intermixing of various castes freely.

dalitvada: a part of the village that is demarcated for the residence of dalits—ex-untouchables—owing to the geographical division of the village on the basis of caste; since these geographical divisions overlap with caste-based divisions they also tend to be subject to the prescribed social rules of touching and not touching that caste implies *vis-à-vis* different groups.

embodied knowledge: the cognitive transmissions and processes that are generated and based on the primacy of the body and its visceral presence in practices that make thinking and knowledge possible in a paradigm that is alternate to written and textual cultures; cognition here functions not only through but also because of the body.

enchantment: a state of wonder emerging from active sensory engagement, often comprising a dissonance of linearity and physicality.

ethnography: in-depth qualitative research arising from immersing oneself in a community to document the lifeways and interactions of its individuals through participant observations, direct interviews and open-ended conversations; while traditionally used in anthropological studies of human communities, its use for understanding nonhuman lifeworlds is slowly gaining momentum.

ethology: the biological study of the behaviour of living entities in their natural environments; given its location in the natural science disciplines, ethology has its own codified set of theories and methods of understanding behaviour, most often in quantitative terms.

expressives: also known as ideophones and mimetics, these words depict vivid sensory perception and mental/emotional states; they are a common part of the lexicon of many languages around the world, but have long been ignored by linguists because of the difficulties in eliciting, analysing and defining them.

extinction: the termination of a kind of organism, usually a species, due to a loss of genetic diversity, excessive predation, population bottlenecks, or adverse environmental change, among other causes; the moment at which the last individual of a species dies; an event common to approximately 99.9 per cent of all species to have lived, including mass extinction events of which there have been five in the earth's history, not including the present rate of extinction which is often classed as the earth's sixth extinction event.

genocide: Article 2 of the 1948 Genocide Treaty broadly defines genocide as political violence directed against a nation to systematically exterminate a human community because of its ethnicity, nationality or religion.

griffin: a newly arrived Englishman in the colony who is unfamiliar with the mores, customs and the practices of the 'natives' as well as the Europeans who have lived in the space for a while.

Hanuman: a prominent deity in Hinduism and a central character in the Indian epic, Ramayana; he is part-human and part-monkey, stands upright, and has the face and long tail typical of a macaque. Hanuman is thought to be fiercely loyal and pious but innovative, adventurous and occasionally mischievous as well.

Harijan: A term brought to popular usage by Mahatma Gandhi, translating into 'children of God', as a way of identifying dalits in the 1930s. However, this term was vehemently opposed by Ambedkar, given its pejorative connotations and since then has fallen into disrepute and been rejected by dalits as a term for self-identification.

hyperobjects: As proposed by Timothy Mortan, hyperobjects are things which are distributed across massive time and space with respect to humans. They cannot be dissociated from human beings and go beyond human temporalities. A hyperobject can be the solar system, biosphere, the sum total of nuclear materials on earth, or human manufactured objects like plastic bags, all of which is beyond the human capacity to control.

individuality: the unique characteristics distinguishing particular individuals from the others of the same group, society or species, shaping their independent behaviours, thoughts, politics or morality, and influenced by their specific sociocultural backgrounds and past experiences; individuality contributes importantly to the variability in responses of human and more-than-human communities to environmental events and phenomena.

interdependence: refers to a reciprocal reliance between human beings, nonhuman animals and other artifacts; from the twin perspective of disability studies and animal studies, it points to the co-dependent relationship between disabled people and service animals. Scholars working at the intersection of these disciplines argue that both parties depend on each other for care, emotional need and effective functioning.

interspecies: the set of relations between species which may take a variety of forms, such as mutualism (cooperative behaviour between species such as bees and pollinating flowers); parasitism (unequal sharing of resources, such as Spanish moss and the southern live oak); hybridity through reproduction (such as the mule as the sterile offspring of a female horse and a male donkey); and a wide array of predator–prey relationships, including predation by humans.

intimacy: a feeling of closeness that binds people as well as beings in the nonhuman worlds; language gives hints regarding the intimate sphere of the language's speakers, covering things such as politeness and respect, kinship, ritual, extending beyond the realm of human society into areas of nature and those that inhabit them.

Long-tailed macaque: a species of macaque, *Macaca fascicularis*, found across south and Southeast Asia, and consisting of nine, morphologically distinct subspecies, of which only the Nicobar long-tailed macaque, *Macaca fascicularis umbrosus*, is found in India; this so-called weed macaque is known for its commensal coexistence with human populations across its distribution range and innovative behaviour strategies that frequently involve direct interactions with humans.

more-than-human: an emerging academic, post-humanistic tradition, where the exceptionalism of human agency in creating society, knowledge and modernity is challenged; it is argued that these are shared achievements and co-constructions by humans and other-than-human agents, both living and nonliving.

Muktijuddho: refers to the nine-month-long civil war that was fought between East and West Pakistan and the international war fought between India and Pakistan, also known as the Bangladesh Liberation War of 1971; the terms 'East Pakistan', 'East Bengal' and Bangladesh are used to denote the names of the states at specific historical periods. The terms 'East Pakistan' and 'East Bengal' were used interchangeably until the birth of Bangladesh in 1971.

ocularcentric: is an adjectival derivation of ocularcentrism, an ideology that privileges sight over other senses; based on the notion that seeing is equivalent to knowing, it posits visual perception as the sole and legitimate means to gain knowledge. Since the blind lack sight, they are perceived as deprived of knowledge, dissociated from 'reality' and incapable of establishing meaningful relationships with others.

Ol: Santali word for 'writing', which includes alphabetic writing but also the drawing of ritual diagrams.

passing: relates to an individual's imitative adoption of the behaviour of a dominant social group; according to sociologist Erving Goffman, it is a strategy that people adopt to conceal their 'true' identities that are stigmatised by political, legal and sociocultural structures. In disability studies, passing refers to an act of a disabled person presenting themselves as able-bodied to meet the expectations of normalcy.

raska: Santali word approximating 'affect', usually used in the context of activities where there is communal enjoyment, such as singing, dancing, eating, drinking, visiting or hosting someone at their home, et cetera.

reading: an act of engaging deeply with a text, not merely visually scanning it for its literal content; reading, in this context, involves a reiterative analysis of the proffered content against the background of the reader's own perspectives, to create a shared understanding of themselves and that of the world around them.

responsibility: Responsibility is instituted when the other issues a call which must be answered by the self. Etymologically, the substance of the response is a promise to respond; its root connotes sacrificial libation which substitutes the subject's lifeblood. The only adequate response, which is to offer oneself, is a promise continually deferred and compensated through endless substitutions.

sanatorium: a medical establishment set up for the treatment and recovery of convalescing patients or individuals suffering from a long-term, chronic ailment; in colonial India, such facilities, especially meant for treating consumptive patients, were located in the hill stations for the want of 'fresh-air' and temperate climate away from the polluting plains.

sensuous landscape: the interrelated networks of sensory perception, affect and morality that are created through communication, representation and play across physical space.

somaesthetics: an interdisciplinary field of inquiry that uses the body as a locus of sensory perception, performance and experience.

Thylacine: *Thylacinus cynocephalus* ('dog-headed pouched-dog') or Tasmanian tiger, a large carnivorous marsupial formerly widespread on the Australian continental landmass as well as the island of New Guinea, and following the last glacial period, common on the island of Tasmania; it is one of two marsupial species in which both sexes possess abdominal pouches, the other being the water opossum of Central and South America. The last living Thylacine specimen is believed to have died in Beaumaris Zoo in Hobart in September 1936, bringing the species to extinction.

vade mecum: a term derived from Latin, meaning to 'go with me'; generally refers to a compact manual, companion, or a pocket-sized guidebook used for ready reference and consultation.

voice: is not only used to emit sounds and locution for communication but is primarily a marker for recognition and identification; by modulating its tone, pitch, volume, an individual voice is employed in parody or mimicry, and over a duration, it prostheticises itself. A person's voice could be broadly divided into three categories: embodied (as sound from the larynx), mental (speech in one's mind), and disembodied (recorded voice).

CONTRIBUTORS

Ajith Kumar formerly taught at the Wildlife Institute of India, Dehradun, and the Salim Ali Centre for Ornithology and Natural History, Coimbatore. He was Director of the postgraduate program in Wildlife Biology and Conservation at the National Centre for Biological Sciences from 2003 to 2020. He began his career in wildlife biology with a survey of primates in southern India in the late 1970s and his doctorate at Cambridge University was on the ecology of the endangered lion-tailed macaque. This study introduced him to the tropical rainforests of the Western Ghats, where his students and he extensively studied herpetofauna, small carnivores, rodents and arboreal mammals for over three decades. His other research interests lie in the Sikkim Himalaya and in marine mammalogy.

Ambika Aiyadurai is currently Assistant Professor (Anthropology) in the Indian Institute of Technology, Gandhinagar. She is an anthropologist of wildlife conservation with a special interest in human–animal relations and community-based conservation projects. Her ongoing and long-term research aims to understand how local and global forces shape human-animal relations. She completed her PhD thesis in Anthropology from the National University of Singapore in 2016. In 2017, she was awarded the Social Sciences Research Council (SSRC) Transregional Research Junior Scholar Fellowship to examine community-based wildlife projects in Northeast India. She was a visiting scholar at the Global Asia Initiative, Duke University in 2018. She is also the author of a monograph titled *Tigers Are Our Brothers: Anthropology of Wildlife Conservation in Northeast India* (2021).

Anindya Sinha, primarily based at the National Institute of Advanced Studies in Bangalore, India, had earlier studied the molecular biochemistry of carbohydrate metabolism in yeast for his doctorate, and subsequently, the social biology of wasps and the classical genetics of human disease for his PhD. His principal research, over the last three decades, has, however, been on behavioural ecology, cognitive ethology, population and behavioural genetics, evolutionary biology and conservation studies of primates and other mammalian species.

His current research in natural philosophies, animal studies, urban ecologies and performance studies primarily concern ethnographic explorations of human–nonhuman relations, especially amidst growing synurbisation, and the lived experiences of non/humans with their promise of unique understandings of more-than-human lifeworlds.

Ankit Kawade is pursuing his MPhil at the Centre for Political Studies, Jawaharlal Nehru University, New Delhi. His research focuses upon the reception of the *Manusmriti* in the writings of Nietzsche and Ambedkar. He is the author of essays such as 'Clearing of the Ground-Ambedkar's Method of Reading' (*CASTE/ A Global Journal on Social Exclusion*, 2021) and 'The Impossibility of Dalit Studies' (*Economic and Political Weekly*, 2019).

Antara Ghatak is Assistant Professor in the Department of English, St. Xavier's University, Kolkata, West Bengal. Antara holds a PhD degree in literature from Calcutta University. She completed her MPhil, MA and BA degrees from University of Calcutta. Her areas of research are Gender Studies and Memory Studies. She also takes interest in studying the representation of women in post-conflict zones.

Antaripa Bharali received her MPhil from the Centre for Political Studies, Jawaharlal Nehru University, New Delhi. Her dissertation is on the political thought of Navayana Buddhism involving a conceptual study of the text, *The Buddha and his Dhamma* by B.R Ambedkar.

Anuparna Mukherjee is currently Assistant Professor of Humanities and Social Sciences, IISER Bhopal. She previously taught in the Department of English at St. Xavier's University, Kolkata. Anuparna holds a PhD in literature from the Australian National University. Anuparna has guest-edited a *Sanglap* special issue on 'City, Space and Literature' with Arunima Bhattacharya. She has published articles in *South Asian Review* ('After the Empire: Narratives of Haunting in the Postcolonial Spectropolis'), *Economic and Political Weekly* and her essay 'Kolkata and the Poetics of Waste in Nabarun Bhattacharya's Spectral City' is part of an anthology titled *Nabarun Bhattacharya: Aesthetics and Politics in a World after Ethics*. She is also in charge of the 'Book review' section of *Sanglap: Journal of Literary and Cultural Inquiry*.

Arka Chattopadhyay is currently Assistant Professor of Humanities and Social Sciences at the Indian Institute of Technology, Gandhinagar. He has contributed to books such as *Deleuze and Beckett, Knots: Post-Lacanian*

Psychoanalysis, Literature and Film, Gerald Murnane: Another World in this One and has been published in journals such as *Textual Practice, Interventions, Samuel Beckett Today/Aujourd'hui, Sound Studies and The Harold Pinter Review.* He has co-edited *Samuel Beckett and the Encounter of Philosophy and Literature* and has guest-edited the SBT/A issue on *Samuel Beckett and the Extensions of the Mind.* He is the founding editor of the online literary journal *Sanglap* and a contributing editor to *Harold Pinter Review.* His first monograph, *Beckett, Lacan and the Mathematical Writing of the Real* was published in 2019. He has recently co-edited a volume on Nabarun Bhattacharya titled *Nabarun Bhattacharya: Aesthetics and Politics in a World after Ethics* (2020) and is working on a monograph on posthumanism, and two edited volumes on affective ecologies and Badiou and modernism.

Asijit Datta is currently Assistant Professor and Head of English at The Heritage College, Calcutta University. In July 2021, he was selected as an Advisory Board member for the School of Liberal Arts and Management Studies at P.P. Savani University, Gujarat. He has also recently started functioning as Assistant Editor at the *Journal of Posthumanism* (Transnational Press, London). He has previously taught at Presidency University, Vidyasagar University, Ramakrishna Mission, Narendrapur, and Bethune College. He received his PhD from the Department of Film Studies, Jadavpur University in 2017. His academic interests pertain to posthumanism, Beckett studies, modern European theatre, world cinema, and psychoanalysis, areas in which he has published several papers and is also an award-winning theatre director and scriptwriter.

Deborah Dutta is a Senior Research Fellow at the Institute of Rural Management, Anand where she has been exploring knowledge and practices pertaining to sustainable agriculture and civic-action initiatives. She is particularly interested in understanding the interplay of community-engaged practices with civic-action initiatives to develop pathways of just sustainability transformations. She received her PhD from the Homi Bhabha Centre for Science Education (HBCSE), where she studied how urban farming initiatives can motivate pro-environmental perspectives. The research involved designing practice-based interventions, particularly community terrace farms, and studying motivation and environmental perspectives emerging from activities in these farms. She likes to blog at https://rhizomesrising.wordpress.com/.

Ishika Ramakrishna is pursuing her PhD at the Centre for Wildlife Studies, Bangalore, India. She has a master's degree in Wildlife Biology and Conservation from the National Centre for Biological Sciences, Bangalore. Her academic and applied interests lie in an interdisciplinary space across ethnoprimatology, social sciences and conservation education. Her current research focuses on the ecological and sociological interface between human and nonhuman primates. Outside of academics, she works extensively with science communication through collaborative enterprise.

Krishna Kumar S is pursuing his PhD from the Department of English Literature, English and Foreign Languages University, Hyderabad, India. His doctoral work focuses on the epistolary practice of blind writers from the early modern period to contemporary times. His research interests include disability studies, life writing, eco-poetics and modern and postmodern poetry. He has participated in a doctoral school titled 'Self (Representing) Disability' conducted by the University of Cologne, Germany in 2019. His article 'Intimacy and the Aesthetic of "Litter" Writing: Epistolary Renarrativisation in Georgina Kleege's *Blind Rage: Letters to Helen Keller*' has been published in *Sanglap: Journal of Literary and Cultural Inquiry*.

Krishnanunni Hari is currently pursuing his PhD from the Department of English Literature, English and Foreign Languages University, Hyderabad. He is working on discourses that figure animals in contemporary India through an analysis of texts ranging from news articles, legal documents and regional film and literature. His research, which attempts to bring critical animal studies and multispecies studies into productive tension with each other, finds its grounding when it is of help to human and nonhuman communities who look for a more just world.

Maan Barua is currently a Lecturer at the University of Cambridge, United Kingdom. He works on the ontologies, economies and politics of the living and material world. His current research focuses on urban ecologies, particularly infrastructure, metabolism and biopolitics. Maan's book *Living Cities: The Urban in A Minor Key* is forthcoming with the University of Minnesota Press and is currently working on a monograph on elephants and plantation life. He is the Principal Investigator on an ERC Horizon 2020 Starting Grant on urban ecologies.

Mark Byron is Associate Professor in the Department of English, University of Sydney. He is author of *Ezra Pound's Eriugena* (2014) and *Samuel Beckett's Geological Imagination* (2020). Along with Sophia Barnes, he has produced the critical manuscript edition of *Ezra Pound's and Olga Rudge's The Blue Spill* (2019). Mark co-edited a dossier with Stefano Rosignoli, 'Samuel Beckett and the Middle Ages', in the *Journal of Beckett Studies* (2016) and is editor of the essay collection *The New Ezra Pound Studies* (2019). He is President of the Ezra Pound Society.

Nathan Badenoch studies the connections between language, culture and ecology in Asia. He was trained in area studies at Kyoto University in Japan and is currently Associate Professor in the Department of Global Interdisciplinary Studies at Villanova University, United States. He has worked in environmental think tanks, bilateral development projects and universities in Laos, Thailand and Japan. His current research interests include poetics and play in language use, changing patterns of multilingualism, and linguistic windows on traditional ecological knowledge. He is involved in ongoing projects to document several endangered languages in Laos, as well as working with adivasi languages in Jharkhand. Recent publications include 'Silence, Cessation and Stasis: The Ethnopoetics of "Silence" in Bit Expressives' in the *Journal of Linguistic Anthropology* (2021), and *Expressives in the South Asia Linguistic Area*, co-edited with Nishaant Choksi.

Nishaant Choksi is currently Assistant Professor of Humanities and Social Sciences at the Indian Institute of Technology, Gandhinagar. His research interests include the study of script, writing, language ideology, education, semiotics, and Indigenous communities in Gujarat, West Bengal, Jharkhand, and Northeast India. He has published in peer-reviewed journals such as *Journal of Linguistic Anthropology, Language and Society, Signs in Society*, and *Modern Asian Studies,* and has also edited and co-edited three volumes, including *Tribal Literature of Gujarat* (2010), *A Course in Mundari* (2015) and *Expressives in the South Asian Linguistic Area* (2020). His monograph, *Graphic Politics in Eastern India: Script and the Quest for Autonomy* was published in 2021.

Prodosh Bhattacharya received his MPhil degree from the Department of English, University of Calcutta. His dissertation was a comparative study of various notions of space in performance of Samuel Beckett's short plays and the works of Badal Sircar, a renowned playwright and theatre personality from Bengal. He has taught as Assistant Professor,

Department of English, St. Xavier's University, Kolkata. At present he is pursuing his doctoral research from the Department of Theatre and Performance Studies, University of Warwick on circus in the context of colonial Bengal, at the intersection of literary and performance studies.

Samrat Sengupta is Assistant Professor and Head of Department of English at Sammilani Mahavidyalaya, University of Calcutta. His research interests include experimental Bengali literature, gender studies, poststructuralism, memory studies and posthumanism. He has contributed a chapter on Bernard Stiegler and Jacques Derrida's take on teletechnology and power in *Critical Posthumanism and Planetary Futures*, edited by Debashish Banerjee and Makarand Paranjpe (2015). He has guest edited a special issue of the journal *Sanglap* on 'Caste in Humanities'. His first Bengali monograph on *Pratibader pathokrom* (*Syllabi of resistance*) has been published in 2022. He is also the co-editor of a volume on Bengali experimental writer Nabarun Bhattacharya titled *Nabarun Bhattacharya: Aesthetics and Politics in a World after Ethics* (2020). He is one of the founding members of the Indian Posthumanism Network.

Sayan Banerjee has trained in both natural sciences and social sciences during his bachelor's and master's degree studies, respectively. He is deeply interested in human–wildlife relations, interdisciplinary conservation science and socioecological studies of Indian forestry. His various projects have documented indigenous hunting in Nagaland, explored gendered implications of human–elephant interactions, and identified the nature and patterns of community participation in wildlife conservation projects, all in northeastern India. He is currently a doctoral student at the National Institute of Advanced Studies in Bangalore, India, investigating behavioural interactions between humans and wild elephants, and their political implications in the non-protected, mixed-use landscapes of Assam.

Toshiki Osada is currently Emeritus Professor, Research Institute for Humanity and Nature, Japan. He received his PhD from Ranchi University, India. His main field of research is linguistic work on the Munda language spoken in Jharkhand, India.

INDEX